T0344716

Robot Learning by Visual
Observation

Robot Learning by Visual Observation

Aleksandar Vakanski
Farrokh Janabi-Sharifi

This edition first published 2017
© 2017 John Wiley & Sons, Inc.

The right of Aleksandar Vakanski and Farrokh Janabi-Sharifi to be identified as the author(s) of this work has been asserted in accordance with law.

Registered Offices
John Wiley & Sons, Inc., 111 River Street, Hoboken, NJ 07030, USA

Editorial Office
111 River Street, Hoboken, NJ 07030, USA

For details of our global editorial offices, customer services, and more information about Wiley products visit us at www.wiley.com.

Wiley also publishes its books in a variety of electronic formats and by print-on-demand. Some content that appears in standard print versions of this book may not be available in other formats.

Library of Congress Cataloguing-in-Publication Data

Names: Vakanski, Aleksandar, author. | Janabi-Sharifi, Farrokh, 1959– author.
Title: Robot learning by visual observation / Aleksandar Vakanski, Farrokh Janabi-Sharifi.
Description: Hoboken, NJ, USA : John Wiley & Sons, Inc., [2017] | Includes bibliographical references and index.
Identifiers: LCCN 2016041712| ISBN 9781119091806 (cloth) | ISBN 9781119091783 | ISBN 9781119091998 (epub) | ISBN 9781119091783 (epdf)
Subjects: LCSH: Robot vision. | Machine learning. | Robots–Control systems. | Automatic programming (Computer science)
Classification: LCC TJ211.3 .V35 2017 | DDC 629.8/92631–dc23
LC record available at https://lccn.loc.gov/2016041712

Cover image: fandijki/Gettyimages

Cover design by Wiley

Set in 10/12pt Warnock by SPi Global, Pondicherry, India

10 9 8 7 6 5 4 3 2 1

To our families

Contents

Preface

The ability to transfer knowledge has played a quintessential role in the advancement of our species. Several evolutionary innovations have significantly leveraged the knowledge transfer. One example is rewiring of the neuronal networks in primates' brains to form the so-called mirror neuron systems, so that when we observe tasks performed by others, a section of the brain that is responsible for observation and a section that is responsible for motor control are concurrently active. Through this, when observing actions, the brain is attempting at the same time to learn how to reproduce these actions. The mirror neuron system represents an especially important learning mechanism among toddlers and young kids, stimulating them to acquire skills by imitating the actions of adults around them. However, the evolutionary processes and modifications are very slow and prodigal, and as we further developed, we tended to rely on employing our creativity in innovating novel means for transferring knowledge. By inventing writing and alphabets as language complements, we were able to record, share, and communicate knowledge at an accelerated rate. Other innovations that followed, such as the printing press, typing machine, television, personal computers, and World Wide Web, each have revolutionized our ability to share knowledge and redefined the foundations for our current level of technological advancement.

As our tools and machines have grown more advanced and sophisticated, society recognized a need to transfer knowledge to the tools in order to improve efficiency and productivity, or to reduce efforts or costs. For instance, in the manufacturing industry, robotic technology has emerged as a principal means in addressing the increased demand for accuracy, speed, and repeatability. However, despite the continuous growth of the number of robotic applications across various domains, the lack of interfaces for quick transfer of knowledge in combination with the lack of intelligence and reasoning abilities has practically limited operations of robots to preprogrammed repetitive tasks performed in structured environments. Robot programming by demonstration (PbD) is a promising form for transferring new skills to robots from observation of skill examples performed by a demonstrator. Borrowed from the observational imitation learning mechanisms among humans, PbD has a potential to reduce the costs for the development of robotic applications in the industry. The intuitive programming style of PbD can allow robot programming by end-users who are experts in performing an industrial task but may not necessarily have programming or technical skills. From a broader perspective, another important motivation for the development of robot PbD systems is the old dream of humankind about robotic assistance in performing everyday domestic tasks. Future advancements in PbD would allow the general population to program domestic and service robots in a natural way by demonstrating the required task in front of a robot learner.

Arguably, robot PbD is currently facing various challenges, and its progress is dependent on the advancements in several other research disciplines. On the other hand, the strong demand for new robotic applications across a wide range of domains, combined with the reduced cost of actuators, sensors, and processing memory, is amounting for unprecedented progress in the field of robotics. Consequently, a major motivation for writing this book is our hope that the next advancements in PbD can further increase the number of robotic applications in the industry and can speed up the advent of robots into our homes and offices for assistance in performing daily tasks.

The book attempts to summarize the recent progress in the robot PbD field. The emphasis is on the approaches for probabilistic learning of tasks at a trajectory level of abstraction. The probabilistic representation of human motions provides a basis for encapsulating relevant information from multiple demonstrated examples of a task. The book presents examples of learning industrial tasks of painting and shot peening by employing hidden Markov models (HMMs) and conditional random fields (CRFs) to probabilistically encode the tasks. Another aspect of robot PbD covered in depth is the integration of vision-based control in PbD systems. The presented methodology for visual learning performs all the steps of a PbD process in the image space of a vision camera. The advantage of such learning approach is the enhanced robustness to modeling and measurement errors.

The book is written at a level that requires a background in robotics and artificial intelligence. Targeted audience consists of researchers and educators in the field, graduate students, undergraduate students with technical knowledge, companies that develop robotic applications, and enthusiasts interested in expanding their knowledge on the topic of robot learning. The reader can benefit from the book by grasping the fundamentals of vision-based learning for robot programming and use the ideas in research and development or educational activities related to robotic technology.

We would like to acknowledge the help of several collaborators and researchers who made the publication of the book possible. We would like to thank Dr. Iraj Mantegh from National Research Council (NRC)—Aerospace Manufacturing Technology Centre (AMTC) in Montréal, Canada, for his valuable contributions toward the presented approaches for robotic learning of industrial tasks using HMMs and CRFs. We are also thankful to Andrew Irish for his collaboration on the aforementioned projects conducted at NRC-Canada. We acknowledge the support from Ryerson University for access to pertinent resources and facilities, and Natural Sciences and Engineering Research Council of Canada (NSERC) for supporting the research presented in the book. We also thank the members of the Robotics, Mechatronics and Automation Laboratory at Ryerson University for their help and support. Particular thanks go to both Dr. Abdul Afram and Dr. Shahir Hasanzadeh who provided useful comments for improving the readability of the book. Last, we would like to express our gratitude to our families for their love, motivation, and encouragement in preparing the book.

List of Abbreviations

CRF	conditional random field
DMPs	dynamic motion primitives
DoFs	degrees of freedom
DTW	dynamic time warping
GMM	Gaussian mixture model
GMR	Gaussian mixture regression
GPR	Gaussian process regression
HMM	hidden Markov model
IBVS	image-based visual servoing
LBG	Linde–Buzo–Gray (algorithm)
PbD	programming by demonstration
PBVS	position-based visual servoing
RMS	root mean square
SE	special Euclidean group

1

Introduction

Robot programming is the specification of the desired motions of the robot such that it may perform sequences of prestored motions or motions computed as functions of sensory input (Lozano-Pérez, 1983).

In today's competitive global economy, shortened life cycles and diversification of the products have pushed the manufacturing industry to adopt more flexible approaches. In the meanwhile, advances in automated flexible manufacturing have made robotic technology an intriguing prospect for small- and medium-sized enterprises (SMEs). However, the complexity of robot programming remains one of the major barriers in adopting robotic technology for SMEs. Moreover, due to the strong competition in the global robot market, historically each of the main robot manufacturers has developed their own proprietary robot software, which further aggravates the matter. As a result, the cost of robotic tasks integration could be many folds of the cost of robot purchase. On the other hand, the applications of robots have gone well beyond the manufacturing to the domains such as household services, where a robot programmer's intervention would be scarce or even impossible. Interaction with robots is increasingly becoming a part of humans' daily activities. Therefore, there is an urgent need for new programming paradigms enabling novice users to program and interact with robots. Among the variety of robot programming approaches, *programming by demonstration* (PbD) holds a great potential to overcome complexities of many programming methods.

This introductory chapter reviews programming approaches and illustrates the position of PbD in the spectrum of robot programming techniques. The PbD architecture is explained next. The chapter continues with applications of PbD and concludes with an outline of the open research problems in PbD.

Robot Learning by Visual Observation, First Edition. Aleksandar Vakanski and Farrokh Janabi-Sharifi.
© 2017 John Wiley & Sons, Inc. Published 2017 by John Wiley & Sons, Inc.

1.1 Robot Programming Methods

A categorization of the robot programming modes based on the taxonomy reported by Biggs and MacDonald (2003) is illustrated in Figure 1.1. The conventional methods for robot programming are classified into manual and automatic, both of which rely heavily on expensive programming expertise for encoding desired robot motions into executable programs.

The *manual programming systems* involve text-based programming and graphical interfaces. In text-based programming, a user develops a program code using either a controller-specific programming language or extensions of a high-level multipurpose language, for example, C++ or Java (Kanayama and Wu, 2000; Hopler and Otter, 2001; Thamma *et al.*, 2004). In both cases, developing the program code is time-consuming and tedious. It requires a robot programming expert and an equipped programming facility, and the outcomes rely on programmer's abilities to successfully encode the required robot performance. Moreover, since robot manufacturers have developed proprietary programming languages, in industrial environments with robots from different manufacturers, programming robots would be even more expensive. The graphical programming systems employ graphs, flowcharts, or diagrams as a medium for creating a program code (Dai and Kampker, 2000; Bischoff *et al.*, 2002). In these systems, low-level robot actions are represented by blocks or icons in a graphical interface. The user creates programs by composing sequences of elementary operations through combination of the graphical units. A subclass of the graphical programming systems is the robotic simulators, which create a virtual model of the robot and the working environment, whereby the virtual robot is employed for emulating the motions of the actual robot (Rooks, 1997). Since the actual robot

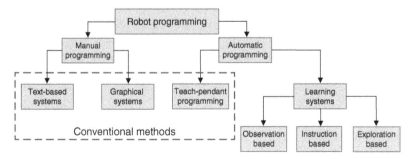

Figure 1.1 Classification of robot programming methods. (Data from Biggs and MacDonald (2003).)

is not utilized during the program development phase, this programming method is referred to as off-line programming (OLP).

The conventional *automatic programming systems* employ a teach-pendant or a panel for guiding the robot links through a set of states to achieve desired goals. The robot's joint positions recorded during the teaching phase are used to create a program code for task execution. Although programming by teach-pendants or panel decreases the level of required expertise, when compared to the text-based programming systems, it still requires trained operators with high technical skills. Other important limitations of the guided programming systems include the difficulties in programming tasks with high accuracy requirements, absence of means for tasks generalizations or for transfer of the generated programs to different robots, etc.

The stated limitations of the conventional programming methods inspired the emergence of a separate class of automatic programming systems, referred to as *learning systems*. The underlying idea of robot learning systems originates from the way we humans acquire new skills and knowledge. Biggs and MacDonald (2003) classified these systems based on the corresponding forms of learning and solving problems in cognitive psychology: exploration, instruction, and observation. In *exploration-based systems*, a robot learns a task with gradually improving the performance by autonomous exploration. These systems are often based on reinforcement learning techniques, which optimize a function of the robot states and actions through assigning rewards for the undertaken actions (Rosenstein and Barto, 2004; Thomaz and Breazeal, 2006; Luger, 2008). *Instructive systems* utilize a sequence of high-level instructions by a human operator for executing preprogrammed robot actions. Gesture-based (Voyles and Khosla, 1999), language-based (Lauria *et al.*, 2002), and multimodal communication (McGuire *et al.*, 2002) approaches have been implemented for programming robots using libraries of primitive robot actions. *Observation-based systems* learn from observation of another agent while executing the task. The PbD paradigm is associated with the observation-based learning systems (Billard *et al.*, 2008).

1.2 Programming by Demonstration

Robot PbD is an important topic in robotics with roots in the way human beings ultimately expect to interact with a robotic system. Robot PbD refers to automatic programming of robots by demonstrating sample tasks and can be viewed as an intuitive way of transferring skill and tasks knowledge to a robot. The term is often used interchangeably with learning by demonstration (LbD) and learning from demonstration (LfD) (Argall *et al.*, 2009; Konidaris *et al.*, 2012). PbD has evolved as an inter-disciplinary field of robotics, human–robot interaction (HRI), sensor fusion, machine learning, machine vision, haptics, and motor control. A few surveys of robot PbD are available in the literature (e.g., Argall *et al.*, 2009). PbD can be perceived as a class of supervised learning problems because the robot learner is presented with a set of labeled training data, and it is required to infer an output function with the capability of generalizing the function to new contexts. In the taxonomy of programming approaches shown in Figure 1.1, PbD is a superior learning-based approach. Compared to the exploration-based learning systems (as an unsupervised learning problem), PbD systems reduce the search space for solutions to a particular task, by relying on the task demonstrations. The learning is also faster because the trial and errors associated with the reinforcement methods are eliminated.

In summary, the main purpose in PbD is to overcome the major obstacles for natural and intuitive way of programming robots, namely lack of programming skills and scarcity of task knowledge. In industrial settings, this translates to reduced time and cost of programming robots by eliminating the involvement of a robot programmer. In interactive robotic platforms, PbD systems can help to better understand the mechanisms of HRI, which is central to social robotics challenges. Moreover, PbD creates a collaborative environment in which humans and robots participate in a teaching/learning process. Hence, PbD can help in developing methods for robot control which integrate safe operation and awareness of the human presence in human–robot collaborative tasks.

1.3 Historical Overview of Robot PbD

Approaches for automatic programming of robots emerged in the 1980s. One of the earlier works was the research by Dufay and Latombe (1984) who implemented inductive learning for the robot assembly tasks of mating two parts. The assembly tasks in this work were described by the geometric models of the parts, and their initial and final relations. Synthesis of program codes in the robotic language was obtained from training and inductive (planning) phases for sets of demonstrated trajectories. In this pioneering work on learning from observation, the sequences of states and actions were represented by flowcharts, where the states described the relations between the mating parts and the sensory conditions.

Another early work on a similar topic is the assembly-plan-from-observation (APO) method (Ikeuchi and Suehiro, 1993). The authors presented a method for learning assembly operations of polyhedral objects. The APO paradigm comprises the following six main steps: temporal segmentation of the observed process into meaningful subtasks, scene objects recognition, recognition of performed assembly task, grasp recognition of the manipulated objects, recognition of the global path of manipulated objects for collision avoidance, and task instantiation for reproducing the observed actions. The contact relations among the manipulated objects and environmental objects were used as a basis for constraining the relative objects movements. Abstract task models were represented by sequences of elementary operations accompanied by sets of relevant parameters (i.e., initial configurations of objects, grasp points, and goal configurations).

Munch *et al.* (1994) elaborated on the role of the teacher as a key element for successful task reproduction. The learning was accomplished through recognition of elementary operations for the observed tasks. The demonstrator supervised and guided the robot's knowledge acquisition by (i) taking into considerations the structure of robot's perceptibility sensors when providing examples, (ii) taking part in preprocessing and segmentation of the demonstrations, and (iii) evaluating the proposed task solution.

Ogawara *et al.* (2002a) proposed to generate a task model from observations of multiple demonstrations of the same task, by extracting particular relationships between the scene objects that are maintained throughout all demonstrations. Each demonstration was represented as a sequence of interactions among the user's hand, a grasped object and the environmental objects. The interactions that were observed in all demonstrations were called essential interactions, whereas the variable parts of the demonstrations called nonessential interactions were ignored in the task planning step. Generalization across multiple demonstrations was carried out by calculating the mean and variance of all trajectories for the essential interactions. A robot program was generated from the mean trajectory, and mapped to robot joints' motors using an inverse kinematics controller.

The advancements in the fields of machine learning and artificial intelligence in the past two decades produced an abundance of new methods and approaches. This trend was reflected by the implementation of approaches in robot PbD based on neural networks (Liu and Asada, 1993; Billard and Hayes, 1999), fuzzy logic (Dillmann *et al.*, 1995), statistical models (Yang *et al.*, 1994; Tso and Liu, 1997; Calinon, 2009), regression techniques (Atkeson *et al.*, 1997; Vijayakumar and Schaal, 2000), etc.

Significant body of work in PbD concentrated on utilizing virtual reality (VR) as an interaction medium to substitute the actual workspace (Takahashi and Sakai, 1991; Aleotti *et al.*, 2004). The main advantages of performing demonstrations in VR include the availability of direct information for the positions and orientations of teacher's motions and the environmental objects during the demonstration phase, the accessible simulation of generated task solutions before the execution in the real world, reduced efforts and fatigue, increased safety of the user compared to physical demonstrations, etc. The concept of virtual fixtures, which refers to the use of virtual guidance and assistance to simplify and improve the tasks demonstration, was employed in Payandeh and Stanisic (2002) via provision of prior task information with an aim to restrict the demonstrated workspace and achieve more consistent performance. Aleotti *et al.* (2004) used visual and tactile virtual fixtures in PbD context, whereas adaptive virtual fixtures that correspond to different subtasks of a complex task were proposed by Aarno *et al.* (2005).

The recent progress in the field of HRI was also taken into consideration by several authors as a basis for improving the process of transfer of knowledge through demonstrations. Since building and developing social mechanisms between robots and humans in a PbD setting rely on successful training, Calinon and Billard (2007a) highlighted several interaction aspects that a demonstrator must reflect on before the demonstrations, such as what are the best ways to convey the knowledge considering the robot's abilities, which parts of the demonstration need special attention, etc. (Figure 1.2).

The latest research in the fields of neuroscience and bioinspiration also stimulated a new stream of research in PbD, and further enhanced its interdisciplinary character. Several researchers drew inspiration from the similar concepts in imitation learning among animals and children, where imitation is not only regarded as a product of social connection, but it also represents an important learning mechanism (Dautenhahn and Nehaniv, 2002).

Today, the PbD paradigm represents a multidisciplinary field which encompasses several research areas. From a general point of view, its goals are to enhance the process of transfer of knowledge to machines by providing motor skill examples through demonstrations.

1.4 PbD System Architecture

The principal steps in solving a typical PbD problem are depicted in Figure 1.3 (Billard *et al.*, 2008). Note that some PbD systems include additional steps (e.g., dashed lines in Figure 1.3), such as evaluation of the

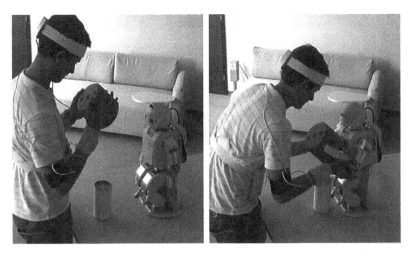

Figure 1.2 The user demonstrates the task in front of a robot learner, and is afterward actively involved in the learning process by moving the robot's arms during the task reproduction attempts to refine the learned skills (Calinon and Billard (2007a). Reproduced with permission of John Benjamins Publishing Company, Amsterdam/ Philadelphia, https://www.benjamins.com/#catalog/journals/is.8.3/main)

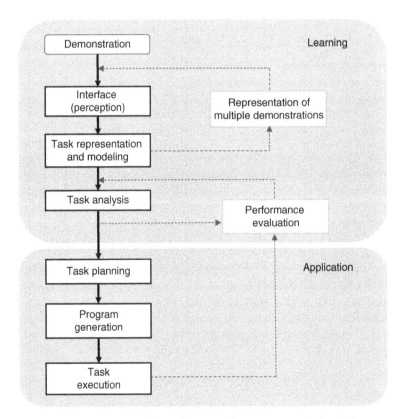

Figure 1.3 Block diagram of the information flow in a general robot PbD system. (Billard *et al.* (2008).)

reproduced task by the end-user and/or provision of additional information for improved robot performance (Chernova and Veloso, 2008a) and simulation of the planned solution before the deployment to robot executive code (Aleotti *et al.*, 2004).

1.4.1 Learning Interfaces

During the observation or perception phase of a PbD process, the teacher demonstrates the task, while the learner observes the teacher's actions and the environment. The learner must have abilities to record the movements of the teacher and the changes in the environment. In other words, the learner agent must possess attributes of perceptibility of actions and states in the world.

The presentation of demonstrations and quality of learning also depends on the learning interface (Figure 1.3). As the PbD advances, it will replace the traditional ways of guiding robots by more user-friendly interfaces, such as sensor-based techniques. The focus of this book will be on sensor-based learning methods. A brief review of all techniques is provided here.

The interfaces used in PbD can be categorized as follows:

- Kinesthetic guidance
- Direct control (through a control panel)
- Teleoperation
- Sensor based (e.g., vision, haptics, force, magnetic, and inertia)
- Virtual reality/augmented reality (VR–AR) environment

In the kinesthetic approach, the robot links are moved manually as shown in Figure 1.4, while in direct control technique the robot links are guided using a provided interface such as control panel. The joint angle or end-point trajectories are recorded and used in the next steps. Alternatively, robots can be controlled using teleoperation, as shown in Figures 1.5 and 1.6. Comparison of nonobservational methods shows that kinesthetic guidance outperforms the other two options in terms of efficiency and effectiveness. However, kinesthetic guidance has usability issues (Fischer *et al.*, 2016). Sensor-based approaches provide ergonomic convenience and ease of task demonstrations. When different sensor-based methods are compared, vision-based observation has the advantage of conveying natural task demonstrations in an unobtrusive way because no sensors are required to be attached to the demonstrator's body or demonstrated objects. Such attachments often lead to increased consciousness on the demonstrator side degrading his/her performance and efficacy.

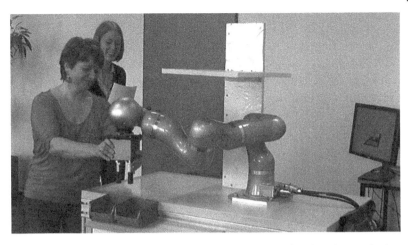

Figure 1.4 Kinesthetic teaching of feasible postures in a confined workspace. During kinesthetic teaching the human operator physically grabs the robot and executes the task. (Seidel *et al.* (2014). Reproduced with permission of IEEE.)

Figure 1.5 The PbD setup for teaching peg-in-hole assembly tasks includes a teleoperated robot gripper and the objects manipulated by a human expert. Tracking is done using magnetic sensors. (Yang *et al.* (2014). Reproduced with permission of IEEE.)

In order to avoid damaging the actual robot, recent approaches also rely on three-dimensional (3D) simulators before executing the program on a real robot. Examples of such platforms include V-REP (Freese *et al.*, 2010) and Gazebo (Koenig and Howard, 2004), which are also open source. The main issue with the simulators is that even if accurate geometric models of

Figure 1.6 Teleoperation scheme for PbD—master arm (on the left) and slave arm (on the right) used for human demonstrations. (Shimizu *et al.* (2013). Reproduced with permission of IEEE.)

the objects become available, the dynamics and control effects are not considered in the simulations, and hence the actual behavior of the robots will deviate from those in the simulations (Angelidis and Vosniakos, 2014). AR, VR, and mixed reality (MR) methods have also been proposed (Fang *et al.*, 2012; Aleotti *et al.*, 2014) to eliminate the need for programming setup. VR–AR techniques bring the advantage of using setups and objects that might not be immediately available or affordable. They also present improved information content because they allow intermittent offline and online programming, enabling a user to modify digital model of the robots while enhancing cognition by adding extra models (AR) or through presenting parts of the real world (MR). AR–VR methods with embedded perceptual/cognitive aids (Figures 1.7 and 1.8) have been shown to outperform working with real robots in training of online programming of industrial robots (Nathanael *et al.*, 2016).

1.4.1.1 Sensor-Based Techniques

Tracking of teacher's hand is an essential observation goal for many tasks, because it is closely related to the grasping states, guiding of the tools, and/or manipulation of the objects of interest. The tracking is usually accomplished by data acquisition from sensing devices mounted directly on the teacher's body (Figure 1.9). In most of the research works on PbD, electromagnetic sensors and data gloves have been used for tracking the teacher's movements (Dillmann, 2004; Martinez and Kragic, 2008). Although these sensing systems are characterized with high measurement accuracy, their operation is sensitive to presence of ferrous parts in the working

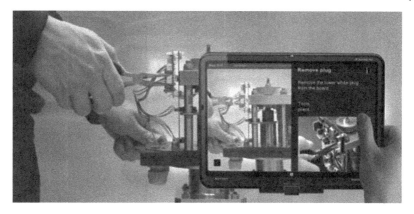

Figure 1.7 AR training of an assembly task using adaptive visual aids (AVAs). (Webel *et al.* (2013). Reproduced with permission of Elsevier.)

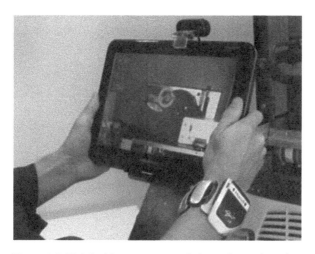

Figure 1.8 Mobile AR component including a haptic bracelet. (Webel *et al.* (2013). Reproduced with permission of Elsevier.)

environment. In addition, the measurement volume of the magnetic trackers is limited, as well as the sensor response becomes nonlinear toward the edges of the measurement volume. PbD interfaces with inertial sensors have also been used for capturing the demonstrator's movements (Calinon, 2009). This type of motion capture device employs gyroscopes for measurement of the rotational rates of the sensors. They are characterized by a large measurement volume, although on the account of reduced accuracy. The main disadvantages of inertial sensory systems

Turn- and
tiltable camera
head for color
and depth
image
processing

Movable platform for
overview perspective

Magnetic field
emitter

Turn- and
tiltable camera
head for color
and depth
image
processing

Loudspeaker
for speech
output and
high sensitive
microphone
for speech
recognition

Data gloves
with attached
tactile sensors
and magnetic
field trackers

Figure 1.9 Sensory systems used for PbD task observation. (Dillmann (2004). Reproduced with permission of Elsevier.)

are the limited positional tracking accuracy, as well as the positional drift that increases over time. Some researchers employed optical marker-based motion capturing for observation of demonstrator's actions in PbD (Kruger *et al.*, 2010). The optical tracking systems are highly accurate, and do not suffer from the interference problems encountered with the magnetic systems. In addition, the measurement volume of the optical systems is large, and it is easily expanded by adding multiple sensors in the working space.

Different from the approaches that involve perception of demonstrations with sensing devices mounted directly on teacher's body, the perception sensors can also be placed externally with respect to the teacher (Argall *et al.*, 2009), as shown in Figure 1.9. This modality for recording of the teacher's motions usually employs vision sensors, for example, a single vision camera (Ogino *et al.*, 2006; Kjellstrom *et al.*, 2008), stereo cameras (Asada *et al.*, 2000), or multiple cameras (Ekvall *et al.*, 2006). The external form of perception is more challenging, due to the difficulties associated with object recognition in cluttered and dynamic environments, determining depths of the scene objects from projections onto the image space, occlusion problems, sensitivity to lighting conditions, etc. On the other hand, this type of perception enables manipulation of scene objects in a natural way without the motion intrusions caused by sensors' wires, and it represents an important step toward the expansion of robotic applications in the service industries.

Fusion of measurements from multiple sensory systems in order to extract the maximum possible information from the demonstrations was proposed by Ehrenmann *et al.* (2001). This work presented an approach for fusing force sensors, a data glove, and an active vision system in a PbD setting, as shown in Figure 1.9. Subsequently, the next generation of intelligent robots must be furnished with efficient techniques for data fusion from multiple sensors, in order to achieve reliable perception of the environment.

1.4.2 Task Representation and Modeling

The learning process in PbD typically relies on the similarities of demonstrated tasks which can be represented in either the *symbolic level* (symbolic encoding) or *trajectory level* (trajectory encoding) (Billard *et al.*, 2008). Hybrid approaches combining trajectory and symbolic learning have also been proposed in the literature, for example, by Ogawara *et al.* (2003). An illustration of different learning levels in PbD is given in Figure 1.10.

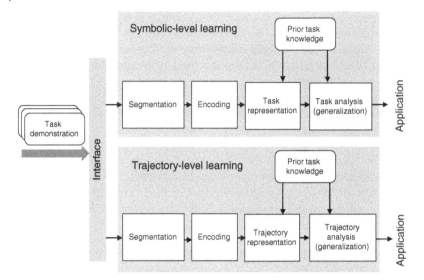

Figure 1.10 Learning levels in PbD.

1.4.2.1 Symbolic Level

Symbolic task representation (Figure 1.10) is based on high-level machine learning methods for skills acquisition as a hierarchical sequence of pre-defined behaviors, referred to as elementary actions or motion primitives (Friedrich *et al.*, 1998; Aramaki *et al.*, 1999; Dillmann, 2004; Saunders *et al.*, 2006). Thus, high-level tasks can be learned through hierarchy rules among a pool of acquired low-level actions (Figure 1.11). Simple elementary actions can be of the form "end-effector moves forward," whereas an example of a goal-directed behavior is "grasp the red box." Symbolic-level learning often uses graph-based or first-order logic for knowledge representation and learning. The encoding involves description of a sequence of known and given primitives. For instance, Nicolescu and Mataric (2001) used a behavior-based network to construct tasks representation, where the nodes in the network represent the behaviors, and the links represent the preconditions and postconditions dependencies. The basic behaviors (i.e., elementary actions) were associated with the state of the environment required for their activation and the predecessor behaviors (preconditions), and also with the effects of the behaviors on the environment and the successor behaviors (postconditions). For instance, the authors evaluated the approach in a task where the goal is to pick up a small box, pass it through a gate formed by a blue and a yellow object in the scene, and drop off the carried box next to an object with orange color. A mobile robot with a manipulator arm was employed for learning

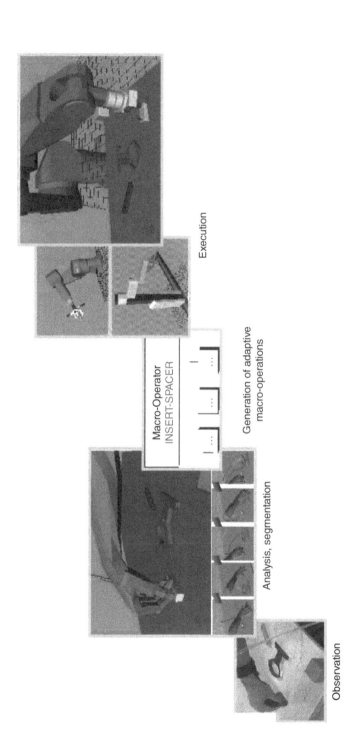

Observation

Analysis, segmentation

Macro-Operator
INSERT-SPACER

Generation of adaptive
macro-operations

Execution

Figure 1.11 Learning at a symbolic level of abstraction by representing the decomposed task into a hierarchy of motion primitives. (Friedrich *et al.* (1998). Reproduced with permission of Springer.)

the task from human demonstrations. The robot learner built the network links based on sequential dependencies of the basic behaviors extracted from the observations. The library of basic behaviors included pick up, track green object, track yellow object, track orange object, and drop. The preconditions for picking up the box require that the drop behavior is active, i.e., the gripper is empty, and that the box is detected. Similarly, postconditions for dropping off the box are that the robot has passed through the gate and has reached the orange object.

The symbolic form of task representation allows predefined behaviors to be reused for different tasks. However, the task representation can fail upon occurrence of a behavior that does not have a corresponding match within the library of preprogrammed behaviors. The main drawbacks of the symbolic task representation are the requirement for efficient segmentation of demonstrations into elementary actions, the requirement to predefine a large set of basic controllers for reproduction of the task sections related to the elementary actions, and their limited applicability with tasks that demand high level of accuracy.

1.4.2.2 Trajectory Level

The trajectory task representation (Figure 1.10) entails encoding of demonstrated actions as continuous signals in the Cartesian space or in the joint angle space. An example of trajectory learning is shown in Figure 1.12. Trajectory-level learning is a low-level learning that relies on the observed trajectories to approximate the underlying demonstrator's policy directly and generalize the trajectories. Representing the demonstrations at a trajectory level is convenient for specifying the velocities and accelerations at different phases of the demonstrated tasks, as well as for defining the spatial task constraints. This type of task representation also allows encoding of arbitrary gestures and motions, conversely to the symbolic-based representation, where the task representation requires prior knowledge about the elementary components that comprise the demonstrated motions. Encoding often involves the use of techniques (i.e., statistical models) to reduce the dimensionality of the segmented signals.

Statistical models can capture the inherently stochastic character of human demonstrated trajectories, and thus many authors employed such models to approximate the demonstrator's policy. These approaches offer compact parametric representation of the observed movements, through probabilistic representation of the variability in the recorded signals. In the work of Calinon and Billard (2008), a Gaussian mixture model (GMM) was exploited for encoding observed tasks, by representing the recorded continuous trajectories as mixtures of Gaussian distributions. Gaussian process regression (GPR) (Schneider and Ertel, 2010) and

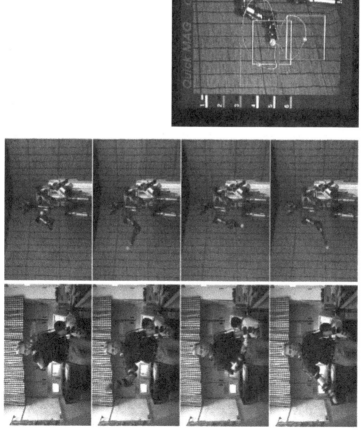

Figure 1.12 A humanoid robot is learning and reproducing trajectories for a figure-8 movement from human demonstrations. (Ijspeert *et al.* (2002b). Reproduced with permission of IEEE.)

Gaussian mixture regression (GMR) (Calinon *et al.*, 2007) have also been used for encoding the trajectories. HMM has been employed by a number of authors (Tso and Liu, 1997; Yang *et al.*, 1997; Aleotti and Caselli, 2006) to model a set of multiple demonstrations, captured as either Cartesian or joint angle trajectories. Due to its robustness to spatiotemporal variations of the observed time sequences, in recent years, HMM has become one of the preferred methods for modeling and analysis of human motions. Alternatively, dynamic systems such as dynamic motion primitives (DMPs) (Ijspeert *et al.*, 2002a) can be utilized where differential equations are introduced to generate 1D movements, and their shapes are approximated by weighted Gaussian basis functions. Such DMPs are used as building blocks for more complex tasks. The main disadvantage of trajectory-level task representation is its inability to reproduce high-level skills.

Initially, PbD had focused on position and velocity trajectories for reproducing efficient task-oriented motions. However, these trajectories might be problematic when tasks involve contact with the environment or a human collaborator. Thus, both force and motion trajectories need to be learned (Kronander and Billard, 2014) for tasks involving compliant motions. The research work on compliant PbD has focused on the following three categories: (i) design of compliant mechanisms with active (e.g., Barret WAM arm), passive (Jafari *et al.*, 2011), or hybrid (Grebenstein *et al.*, 2011) robot joints to provide adaptive robot joint impedance for safe implementation of PbD; (ii) online tuning of impedance for given tasks using accurate model of robot–environment interactions (Chan and Liaw, 1996); and (iii) impedance adaptation skills learning through demonstrations of human's adaptive compliance (Ajoudani *et al.*, 2012) to enable variation of robot (joint or task) stiffness during physical interactions with a human collaborator. Sensory data from electromyography (EMG) signals (Peternel *et al.*, 2014) and haptic information (Kronander and Billard, 2014) can be used for learning the impedance regulation.

1.4.3 Task Analysis and Planning

The task analysis and planning phases in PbD consist of establishing a mapping between the demonstrated actions and the corresponding actions the robot learner should undertake to achieve the task goals.

1.4.3.1 Symbolic Level
For tasks represented at a symbolic level, the task analysis is often related to recognition of the relevant relationships between the teacher's hand, the manipulated objects, and the environmental objects. Hence, it is important in this phase to establish relations between the states of the

environment and the corresponding actions, for example, conclusions about which states and actions must occur before other states. The task analysis can also involve elimination of suboptimal demonstrations, extraction of task constraints, aggregation of the demonstrations into different patterns, etc.

The task planning refers to generating a sequence of robot actions for attaining the desired states. It involves interpretation of the spatial relationships of the relevant objects and generation of an appropriate sequential behavior that will accomplish the task goals without violation of the task constraints. For instance, in the behavior-based network architecture reported by Nicolescu and Mataric (2001), the basic predefined behaviors were activated when all the preconditions (observed states in the environment) and the postconditions (action effects) were satisfied. Similarly, in the article by Ekvall and Kragic (2008), a variant of the Stanford Research Institute Problem Solver (STRIPS) planner was employed for teaching a robot the essential order of the subtasks for accomplishing the task goals.

1.4.3.2 Trajectory Level

The task analysis and planning in trajectory-based approaches imposes generation of a trajectory for accomplishing the task goals. A body of literature applied imitation of a single demonstrated trajectory by a robot learner (Asada *et al.*, 2000). However, these frameworks for transfer of knowledge do not provide cognitive abilities to the learning system. A type of learning system that possesses advanced cognitive abilities relies on learning from multiple demonstrated trajectories and selecting a single trajectory from the demonstrated set, which is the most adequate for reproducing the task objectives (Tso and Liu, 1997; Calinon and Billard, 2004). The drawback of these approaches is that the reproduction trajectory is retrieved from only one of the demonstrated trajectories, which is selected as the most consistent across the demonstrated set. The PbD systems with a higher level of cognition are capable of generalizing from a set of repetitive demonstrations in creating a trajectory for task reproduction.

The step of planning a reproduction strategy at a trajectory level of representation is often based on regression techniques. For instance, the work by Calinon *et al.* (2007) used GMR to obtain a generalized version of tasks modeled by GMM. This method produces smooth generalized trajectories by taking into consideration the covariances of the Gaussian distributions, and is also suitable for representing the spatial task constraints based on the variance across the demonstrated motions. Third-order spline regression (Calinon and Billard, 2004) and nonuniform rational B-splines (NURBS) (Aleotti and Caselli, 2005) have also been used for creating a generalized trajectory from sets of extracted key points in the observations (Inamura *et al.*, 2003) based on statistical task modeling with HMM. Atkeson *et al.* (1997) proposed to use locally weighted regression for skill acquisition, whereas Schaal and Atkeson (1998) employed receptive field weighted regression in the context of incremental learning of a fitting function.

In addition, several works employed the theory of dynamical systems for reproduction of trajectories modeled as a critically damped mass-spring-damper system (Ijspeert *et al.*, 2003; Gribovskaya *et al.*, 2010). The advantage of such systems is the independence of explicit time indexing in creating a reproduction strategy, in a sense that the system dynamics can evolve toward achieving a discrete goal or toward maintaining a periodic motion (Ijspeert *et al.*, 2002b) without temporal dependence on the demonstrated data (Figure 1.12).

1.4.4 Program Generation and Task Execution

In the program generation step, the problem solution strategy from the task planning is translated into an executable robot program, which is afterward transferred to the robotic platform for execution of the desired motions.

For tasks represented by a sequence of symbolic cues, the planned actions for achieving the task goals are mapped onto a repertoire of pre-programmed robot primitives, i.e., elementary actions (Nicolescu and Mataric, 2001; Dillmann, 2004). The resulting program code in the native robot language is deployed on the robot learner platform and the planned sequence of elementary actions is executed in the actual environment. Translation of the generated plans for task reproduction onto a different robotic platform is straightforward, provided that the required elementary actions are predefined with the other robot, and it supplies the basic capabilities for achieving the task goals, such as degrees of freedom (DoFs) and workspace.

Generating a program for tasks represented at a trajectory level largely depends on the used parameters for generation of reproduction strategy. For plans described with Cartesian poses of robot's end-effector or manipulated objects, the inverse kinematics problem is solved for calculating the robot joint angles, which are sent as command signals to the robot controller. In fact, most often an inverse differential kinematics algorithm is used, since these algorithms provide a linear mapping between the joint space variables and the operational space variables. Examples are the least norm (i.e., pseudoinverse) method (Whitney, 1969), weighted least-norm method (Whitney, 1972), damped least-squares method (Nakamura and Hanafusa, 1986; Wampler, 1986), etc. On the other hand, in some PbD approaches, the joint angles of the demonstrator's hand are recorded, and the planned strategy for task reproduction is expressed in the desired joint angles for the robot learner. In that case, the program generation may be preceded by scaling of the joint angles trajectories, to accommodate for the different kinematic parameters between the teacher and learner agents (Calinon *et al.*, 2005). The task reproduction step involves deployment of the executable program onto the robot's low-level controller and execution of the program.

Some PbD systems endow validation of the generated robot programs via a simulation phase before the actual execution by the robot. For instance, in the work of Ehrenmann *et al.* (2002), the task abstraction and the generated reproduction plans were shown on a graphical interface, and a human supervisor was prompted to accept or reject the proposed sequence of actions. Only after the approval by the end-user, the code was transferred to a robot and the task was executed in the real-world conditions.

1.5 Applications

From application point of view, PbD will not be limited to programming industrial manipulators. An important growing area of application is service robotics, where the service recipients are often novice users, for example, use of homecare robots by the elderly. Entertainment and security robotics are other important areas of growth, where a PbD capability will extend quality and quantity of such applications. PbD has also been extended for use in humanoid robots.

Various applications of PbD under development have been summarized as follows:

Industrial robots—These robots are usually referred to as robotic manipulators with special-purpose end-effector. However, automated guided vehicles (AGVs) can also be treated as industrial robots. The working environment of industrial robots is mostly structured. The main industrial application is for manufacturing automation across a variety of applications including material handling, assembly, welding, dispensing, processing (e.g., sealing, cutting, casting, and surface finishing), and inspection. Due to the increased demand for flexibility and reduced structured environments, programming costs of industrial robots contribute significantly to the production costs. Therefore, PbD could play a major role in reducing the time and cost of production. Examples include robotics assembly and surface finishing.

1) *Robotic assembly*—PbD has been used for teaching difficult task of peg-in-hole for assembly operations (Yang *et al.*, 2014). In new directions of manufacturing (i.e., just-in-time manufacturing), the assembly

operations need to be highly flexible and adaptive to many tasks. Additionally, programming of small parts assembly is difficult as the clearance may exceed the robot motion accuracy. The robot must master different control modes from free to compliant motions and their coordination. Therefore, to lower the cost and time of programming, the use of PbD is very encouraging.

2) *Robotic surface finishing*—Variations in the objects geometry, surface roughness, and required tolerances require highly skilled operation knowledge to adapt to such changes and adjust tool paths and parameters. Subsequently, research is underway to use PbD to capture suitable finishing parameters and to adjust industrial robots tool path for a given surface finishing task (Ng *et al.*, 2014).

Humanoid robots—Large number of DoFs and kinematic redundancies in humanoid robots make their programming quite challenging (Figure 1.13). Therefore, PbD approaches have been adopted to teach various tasks such as visuomotor coordination (Lemme *et al.*, 2013), and natural interactions with a human by a social robot (Liu *et al.*, 2014).

Service robots—Robots are expected to work closely with humans to help them in their daily lives. An important feature of these robots is their close interaction and even collaborations with a human being. For this purpose, a common understanding of the collaborative tasks is a key element of the future service robots (Rozo *et al.*, 2016). While in some scenarios, such collaboration is meant to increase process efficiency, in many tasks such as those in houses and hospitals, physical contact is necessary. Additionally, such robots will face many variations in their scenes on a day-to-day basis. Many users of service robots will also lack programming skills. Obviously, conventional programming approaches cannot be adopted for service robots. Alternatively, PbD approaches hold great potential for ease of programming through capturing the task models by demonstrations. One such example is shown in Figure 1.14.

Medical robots—Many medical procedures require high dexterity and specialized skills which are usually acquired through intensive training and pose high financial burden to healthcare systems. In the meanwhile, many medical procedures follow similar steps and tasks. Besides, in the

Figure 1.13 Control of a 19 DoFs humanoid robot using PbD. (Field *et al.* (2016). Reproduced with permission of IEEE.)

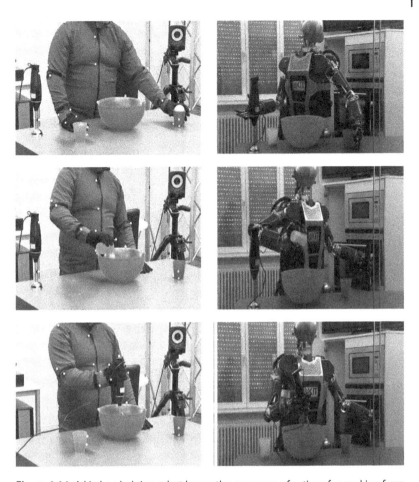

Figure 1.14 A kitchen helping robot learns the sequence of actions for cooking from observation of human demonstrations. (Wachter *et al.* (2013). Reproduced with permission of IEEE.)

absence of a specialist, robotic intervention could prove vital. Therefore, medical robotics are good candidates for integrating PbD (van den Berg *et al.*, 2010). For instance, the extended focused assessment with sonography in trauma (eFAST) has been proved very effective in identifying the internal bleedings in prehospital settings like ambulances. Despite its advantages such as portability and noninvasiveness, the lack of experienced sonography operators has limited its applications. In Mylonas *et al.* (2013), a lightweight robot is programmed by expert demonstrations for eFAST scanning and is shown in Figure 1.15.

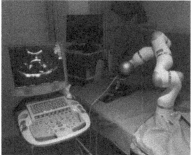

Figure 1.15 The experimental setup used for teaching (on the left) includes an ultrasound machine, an ultrasound phantom model and a handheld ultrasound transducer with force sensing and built-in 3D position markers for optical tracking system. The robotically controlled ultrasound scanning is also shown (on the right). (Mylonas *et al.* (2013). Reproduced with permission of IEEE.)

In addition to motion trajectory and task learning in the aforementioned domains of applications, PbD platform can be used for control and planning purposes as well. From this perspective, typical applications include motion planning, grasp planning, and compliance planning.

Social robots—The presence of robots in human populated environments will inevitably increase in the years to come, thereby dictating the development of complex interactions between robots and humans. PbD can provide a platform for learning and building social mechanisms between humans and machines. For instance, transferring knowledge to a robot can also be perceived from the educational point of view. That is, such transfer of knowledge would have similar effects as the transfer of knowledge to the children or other humans, and this can stimulate interest in the humans to observe and support the learning progress by the robot. Thus, such interactions can cause emotional involvement in the interaction with the robot learner (Calinon and Billard, 2007a).

Robot motion planning—With increased number of DoFs and presence of obstacles in human–robot collaborative workspaces, motion planning problem could become quite challenging. In particular, some of the obstacles might change from each task to another task (Seidel *et al.*, 2014). PbD approach can be used to create incrementally a graph-based representation of the demonstrated obstacle-free task space which can be utilized for planning of safe paths. A real-time trajectory modification algorithm can also be integrated to change the learned trajectory by PbD when additional task constraints such as short distance and obstacle avoidance constraints are introduced (Kim *et al.*, 2015).

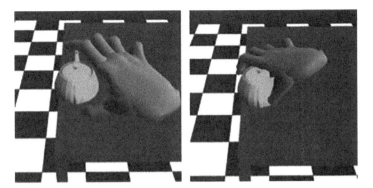

Figure 1.16 Robot grasp planning application. (Aleotti and Caselli (2010). Reproduced with permission of Elsevier.)

Robot grasp planning—Planning quality grasps for given objects and tasks with complex robotic hands could be quite challenging. PbD can be used to plan effective human-like grasps (Figure 1.16). After recognition of contact states and each finger approach direction taught by an expert as the basis of understanding grasp, it can be mapped onto a robot gripper device which is usually very different from human hand. Task knowledge is then applied via high-level reasoning to choose among the applicable grasps (Aleotti and Caselli, 2010).

Robot compliance planning (*impedance learning*)—In addition to motion trajectory learning, impedance learning can enable robots to reproduce many collaborative tasks and compliant motions with the environment. Various sensors or combinations of them can be used. For example, EMG signals can be used to teach various compliance levels to a robot for a given task (Peternel *et al.*, 2014). In another work (Kronander and Billard, 2014), a teacher shook and also firmly held the robot and the haptic information was used to teach when low and high stiffness gains were required.

1.6 Research Challenges

Despite the significant advances in PbD and robotic technology, there are open issues that remain to be addressed. At the present time, transfer of skills to robots with PbD poses many challenges. In order to learn from demonstrations, the robotic system must possess certain level of cognitive skills. It must be able to reliably perceive the states and actions of the environment, to create an abstract representation of demonstrated tasks, and to generate a plan for reproduction of the demonstrated actions. Furthermore, the learning system should have the ability of generalizing the observed task solutions across different initial arrangements of the scene objects. Examples of open issues include the following:

- Creating robot programs from demonstrations for generic tasks
- Correct interpretation of the intent of the demonstrator when accomplishing a task
- Compensating for the differences in kinematic and dynamic parameters, and differences in the DoFs when the teacher and the learner have dissimilar bodies
- Relating the model of the human kinematics to task learning and reproduction
- Dealing effectively with nonoptimal and ambiguous demonstrations, or with the failures in executing the task
- Developing robust learning methods for achieving the desired performance under disturbances and changes in the environment
- Evaluating the performance of the robot in skills acquisition, and the performance of the teacher in skills transfer
- Provision of prior knowledge in a balanced way to speed up learning

The remaining text in this section discusses several of the above research challenges in robot PbD.

1.6.1 Extracting the Teacher's Intention from Observations

Certainly, one crucial element in learning from observation of demonstrations is the interpretation of the teacher's intention by the robot learner. One aspect of it is to interpret the interactions among the subtasks, that is, to determine for which of the subtasks the order of execution is important, and for which subtasks it is not. For example, it is important to know whether achieving the final goal of the task is sufficient for successful task reproduction (e.g., peg-in-hole task), or whether the task requires achieving many subgoals (e.g., trajectory following for a welding task). With regard to these questions, most of the studies in PbD relied on one of the following: (i) the teacher instructs the robot about the possible intentions during the demonstrations, or (ii) the teacher's intentions are extracted from multiple observations of demonstrations.

An example of the first set is the work of Wu and Kofman (2008) who proposed the teacher to give a brief task description to the learner before the demonstration phase. The description provided information about the overall structure of the task and its subgoals, which helped the learner in creating the task hierarchy before the demonstrations. In addition, during the demonstrations, the teacher provided short voice descriptions in the transitions between the subtasks, which alleviated the task segmentation by the robot. Several works proposed to use pointing and gazing cues during the demonstrations (Scassellati, 1999; Calinon and Billard, 2006). In that scenario, the teacher gazes toward the object of interest and/or points with his hand toward the object, so that the learner can infer the intents of the teacher's actions.

The second type of approaches revolves around extracting the teacher's intention from multiple demonstrations of the same skill. For instance, the methods for robot PbD using statistical models as a mathematical tool for task representation (e.g., HMM) employ latent states in describing the intention of the teacher during the demonstrations. The mental states of the teacher, that is, his/her intentions in performing particular tasks, are observed through a measurable process, which are the teacher's executed actions (Yang *et al.*, 1997). The learner observes the demonstrations and attempts to extract the teacher's intentions from the constraints on the demonstrations described in a probabilistic framework. In the work of Calinon (2009), for demonstrations encoded with mixtures of Gaussians, the covariance matrix of the corresponding Gaussian distribution for each segment of the demonstrations was used to induce the task constraints. The segments of the task with small variability across demonstrations were associated with highly constrained motions. Thus, the learner could infer whether the teacher intended to perform precise or loose movements for the different parts of the trajectories.

The research work in the book concentrates on extracting the teacher's intentions from multiple demonstrations of the same task. Hence, the robot learner should possess abilities to identify the relevant task constraints and to extract task-specific knowledge from the observed task examples. Although augmenting the robot's task knowledge with teacher's instructions is also beneficial and leverages the learning, the topic is beyond the scope of the book.

1.6.2 Robust Learning from Observations

1.6.2.1 Robust Encoding of Demonstrated Motions

Robust encoding of human demonstrations relates to the selection of a set of training examples that conveys quantitative and qualitative information about a skill, which is sufficient for deriving a model of the skill. Efficient learning implies fast skill transfer from a few examples, since providing a large number of demonstrations for the same task can be irritating for the demonstrator, and it can cause fatigue and poor demonstrating performance. However, a limited number of demonstrations can lead to undemonstrated portions of the task. Therefore, an important question in robot PbD is the extent of training examples, i.e., how many demonstrations are enough to teach a skill. Within the PbD literature, generally between four and six demonstrations are employed for skill transfer to robots, although in some cases greater number of demonstrations are exploited (e.g., 10 demonstrations in the approach of Martinez and Kragic (2008) and 25 demonstrations in the work of Gams and Ude (2009)).

In addition, the approaches for robust learning should provide a means for dealing with ambiguity and suboptimality in the demonstrations. The demonstrated dataset can be ambiguous with respect to the possibilities for the teacher to achieve the same effect on the environment with different actions. Another source of possible ambiguities relates to the difficulties of the robot learner to distinguish two similar but different tasks, based on the recorded data with the available sensory system. Furthermore, the demonstrated set can be suboptimal in a sense that the individual demonstrations differ significantly, and the robot learner is not able to generate a successful reproduction plan.

The most intuitive approach for dealing with suboptimal and ambiguous demonstrations is to let the teacher select the demonstrated examples that will transmit enough information to the learner about the demonstrated task. This approach assumes that the teacher would speed up the learning by removing the nonimportant demonstration examples, using his/her natural abilities for generalization. Some authors proposed learning from demonstrations performed by several teachers, which can reduce the suboptimality of the performance by the individual teachers (Pook and Ballard, 1993). In the approach proposed by Chernova and Veloso (2008b) if the robot is not confident about some elements of the demonstrated task, it requests the teacher to perform additional demonstration(s). This way the robot incrementally builds the task model until the level of confidence for the entire task is above a certain threshold value.

Still, it is not clear how to automate the process of selection of training examples without relying on the generalization abilities of the teacher. Aleotti and Caselli (2006) proposed a distance metric for clustering the demonstrations into similar patterns, so that the clusters of trajectories can be treated as different skills. This approach can fail for sets containing large temporal variations across the demonstrations, or for sets with sparse solution space. On the other hand, enhanced robustness of the learning process can be achieved by eliminating the demonstrations that are too dissimilar to the set. Namely, once an appropriate model of the skill is created, the likelihood that a training example is generated by the learned model can be employed as a metric for deciding whether to use that specific training example for further processing and learning (Ogawara *et al.*, 2002b). Nevertheless, the model of the skill is only as good as the demonstrated examples so that deriving a model from a set of suboptimal demonstrations can result in a suboptimal model.

1.6.2.2 Robust Reproduction of PbD Plans

Another challenge in PbD is designing a controller that will ensure robust execution of the planned strategy for task reproduction.

In general, most of the PbD methods rely on strict temporal ordering of the states in the reproduction strategy. Hence, during the execution of the planned strategy, the robot attempts to attain the specified sequence of joint configurations, employing the explicit temporal ordering. The disadvantage of these PbD systems is the lack of flexibility in the task execution, that is, the systems lack robustness to perturbations and deviations from the ideal task configurations.

On the other hand, the design of robust controllers that will achieve the task goals in presence of perturbations (i.e., modeling errors and measurement noise) and changes in the environment is an open question in the PbD field. Within the literature, a body of work employed the dynamical systems approach (Ijspeert *et al.*, 2003) for generating reproduction strategies with enhanced robustness properties. For tasks where the only important goal is to attain a particular state at the end of the movement (so-called discrete tasks), the dynamical systems approaches ensure convergence toward the final state in presence of perturbations. However, for complex tasks (that involve accomplishing several subgoals), it is challenging to develop controllers that will execute under disturbances. Such advanced controllers should be able to perform replanning in real time by devising an alternative control strategy for achieving the desired goals (Gribovskaya *et al.*, 2010). Under significant level of perturbations that can cause large deviations from the initial strategy for task reproduction, it may be unclear for the robot how to proceed with the task execution in the new situation. The replanned strategy may require reconsidering the relative importance for execution of the different components of the task, and reformulation of the task constraints for obstacle avoidance, robot workspace limitations, etc.

1.6.3 Metrics for Evaluation of Learned Skills

The field of robot PbD currently lacks criteria for evaluation of the learning performance of the system. The difficulties arise from the fact that most of the studies in the literature focus on solving specific robotic applications, using different levels of task abstraction and task representation. The development of comprehensive evaluation metrics for the robot learning will enable comparison between the different approaches and application domains, and will provide a basis for solutions to generic tasks. This section briefly overviews several proposed techniques for evaluation of the skill acquisition in trajectory-based learning.

Pomplun and Mataric (2000) proposed a set of metrics based on the mean-square difference between the corresponding joint angles in the demonstrated and reproduced movements. The results showed that the metric based on segmented trajectories into sequences of high-level movements yields the most satisfactory results. Calinon and Billard (2007b) argued that assessment of the imitation performance should put more weight on those dimensions of the trajectories that are more constrained across the demonstrations, and thus are more important for successful reproduction. Calinon *et al.* (2005) introduced a metric that encapsulates goal-directed task constraints, that is, it enables the reproduced performance by the robot to be evaluated with a combination of functions involving distance metrics and task goals (Billard *et al.*, 2004). The method introduced a set of weighting factors for the level of importance of each of the several cost functions; however, the weights were manually selected by the demonstrator based on his understanding of the task goals.

A set of metrics for evaluation of the HRI in PbD was presented by Steinfeld *et al.* (2006). The proposed metrics pertain to assessment of the individual and joint efforts of a human and a robot in performing a task. Accordingly, the robot performance was evaluated through the degree of autonomous operation, abilities of self-assessment, awareness of human's presence, etc. The human operator performance was rated based on his/her level of knowledge of the task, mental abilities, knowledge of robot's abilities, etc. Evaluation of the performance of the human–robot as a team involved: effectiveness (percentage of the task that was performed with the designed autonomy); time efficiency; rate of utilization (e.g., percentage of request for help made by the robot, or by the human); etc. The authors also listed a number of biasing factors that have effects on the evaluation and many task-specific metrics.

1.6.4 Correspondence Problem

The process of skill transfer in PbD imposes the correspondence problem between the agent embodiments. In situations where both the teacher and the learner are robots with identical body structure, the transfer of knowledge infers direct mapping of the corresponding movements. On the other hand, learning of skills from demonstrations performed by a human teacher (or robots with different structures) involves solving the correspondence problems of workspace constraints, different DoFs, different kinematic and dynamic characteristics, etc.

Nehaniv and Dautenhahn (2001) examined the correspondence problem for a general case of imitation, which can be applied to biological or artificial agents. Their work developed a mathematical approach to model the mapping in task imitation, using the concept of relational homomorphisms. The level of success in imitation was exploited through the effects on the environment, in a sense that a successful imitation refers to accomplishing a desired effect using the affordances of the agent–environment coupling. Starting from the fact that a demonstrator defines the goal of the task, the authors held that the evaluation of the task reproduction is also demonstrator-dependent.

Alissandrakis *et al.* (2007) presented linear correspondence matrices for describing the mapping of the different DoFs between the teacher and the learner. State and action metrics for evaluation of the imitation matching were introduced, with the imitation process aiming to minimize the values of the states and action metrics. However, the proposed solutions based on linear mapping of the corresponding DoFs have limited applicability. In fact, in a PbD environment with a human teacher and a robot learner, the embodiment mapping is nonlinear and highly complex.

The research presented in the book is not concentrated on solving the correspondence problem in PbD, which nevertheless remains one of the open areas for future research.

1.6.5   Role of the Teacher in PbD

The level of HRI varies in different robot programming systems. For instance, traditional robot programming systems based on manually writing the program codes do not require interaction between the user and the robot, whereas the robot programming systems based on guiding the robot's links through required trajectories with a teach-pendant involve higher level of interaction. The HRI in PbD is more intense, and it can affect the speed of skill acquisition by the robot. Hence, one of the open questions in the PbD learning relates to the role of the demonstrator.

Several authors suggested that the human teacher should play an active role in the process of skill transfer to a robot. Starting from the fact that in transferring knowledge among humans both the teacher and the learner have active roles, the same concept can be implemented for the robot PbD problem by putting the teacher in the loop of learning. For instance, in the work of Friedrich *et al.* (1998), the teacher overviews the entire learning process, and either approves the execution of the final program code or rejects the proposed code and re-explains some parts of the task.

Calinon and Billard (2007a) suggested that both the teacher and the learner should have active roles not only during the demonstration and learning phases but also during the task execution phase. Namely, when the teacher observes the robot executing the task, he/she recognizes which parts of the task were not performed at a satisfactory level. Subsequently, the teacher would help the learner to improve the task reproduction by providing feedback and refining the robot performance. The same work also reviewed the physiological and sociological aspects of teaching robots and evaluation of the reproduction attempts. Accordingly, the degree of HRI during the teaching can be evaluated based on the human's involvement in the interaction and the enthusiasm in transferring knowledge. However, the quantification of teacher's involvement/enthusiasm in the teaching interaction is even more challenging in comparison to the evaluation of learner's ability to encode skills and to create a generalized version of the skill from a set of demonstrations. Using insights from pedagogy and developmental sciences, the authors suggested several benchmarks for evaluating the success of the knowledge transfer in PbD scenarios. Furthermore, it was argued that the teacher should be knowledgeable about the learner's abilities, in terms of the ways of learning skills, but also the range of motions, velocities limits, and similar characteristics of the learning agent. Consequently, he/she should adapt the teaching techniques to maximize the probability of fast and proper transfer of knowledge.

Investigation of the social mechanisms in HRI is not among the objectives of the book. However, with the increased presence of robots around us and the increased number of services that will be provided by robots in the near future, it will be inevitable that we learn how to interact with the robots in effective and safe ways. Many research resources are currently devoted to this topic outside the PbD field.

1.7 Summary

The chapter provides an introduction to robot PbD. The motivations for the development of PbD systems along with an overview of the existing literature related to the PbD paradigm are presented. Several important early works and approaches from the PbD domain are reviewed, followed by discussion on recent advancements and state of the art. The main steps in solving the PbD problem are classified into perception, task representation and modeling, task analysis and planning, program generation, and task execution. An overview of the methods for solving the individual phases of PbD is presented for task represented at both symbolic and trajectory level of abstraction. Examples of PbD applications in the domains of industrial and service robotics are presented. The major challenges and open questions within the robot PbD field are discussed in the last section of the chapter.

References

Aarno, D., Ekvall, S., and Kragic, D., (2005). Adaptive virtual fixtures for machine-assisted teleoperation tasks. *Proceedings of IEEE International Conference on Robotics and Automation*, Barcelona, Spain, pp. 1139–1144.

Ajoudani, A., Gabiccini, M., Tsagarakis, N., Albu-Schaffer, A., and Bicchi, A., (2012). Teleimpedance: exploring the role of common-mode and configuration-dependent stiffness. *Proceedings of 2012 IEEE-RAS International Conference on Humanoid Robots*, Osaka, Japan, pp. 363–369.

Aleotti, J., and Caselli, S., (2005). Trajectory clustering and stochastic approximation for robot programming by demonstration. *Proceedings of IEEE/RSJ International Conference on Intelligent Robots and Systems*, Edmonton, Canada, pp. 1029–1034.

Aleotti, J., and Caselli, S., (2006). Robust trajectory learning and approximation for robot programming by demonstration. *Robotics and Autonomous Systems*, vol. **54**, no. 5, pp. 409–413.

Aleotti, J., and Caselli, S., (2010). Interactive teaching of task-oriented robot grasps. *Robotics and Autonomous Systems*, vol. **58**, pp. 539–550.

Aleotti, J., Caselli, S., and Regiani, M., (2004). Leveraging on a virtual environment for robot programming by demonstration. *Robotics and Autonomous Systems*, vol. **47**, no. 2–3, pp. 153–161.

Aleotti, J., Micconi, G., and Caselli, S., (2014). Programming manipulating tasks by demonstration in visuo-haptic augmented reality. *Proceedings of 2014 IEEE International Symposium on Haptic, Audio and Visual Environments and Games*, Richardson, USA, pp. 13–18.

Alissandrakis, A., Nehaniv, C.L., and Duatenhahn, K., (2007). Correspondence mapping induced state and action metrics for robotic imitation. *IEEE Transactions on Systems, Man, and Cybernetics—Part B*, vol. **37**, no. 2, pp. 299–307.

Angelidis, A., and Vosniakos, G.-C., (2014). Prediction and compensation of relative position error along industrial robot end-effector paths. *International Journal of Precision Engineering Manufacturing*, vol. **15**, no. 1, pp. 66–73.

Aramaki, S., Nagasawa, I., Kurono, S., Nagai, T., and Suetsugu, T., (1999). The knowledge representation and programming for robotic assembly task. *Proceedings of IEEE International Symposium on Assembly and Task Planning*, Porto, Portugal, pp. 256–261.

Argall, B., Chernova, S., Veloso, M., and Browning, B., (2009). A survey of learning from demonstration. *Robotics and Autonomous Systems*, vol. **57**, no. 5, pp. 469–483.

Asada, M., Yoshikawa, Y., and Hososda, K., (2000). Learning by observation without three-dimensional reconstruction. *Proceedings of Sixth International Conference on Intelligent Autonomous Systems*, Venice, Italy, pp. 555–560.

Atkeson, C.G., Moore, A.W., and Schaal, S., (1997). Locally weighted learning for control. *Artificial Intelligence Review*, vol. **11**, no. 1–5, pp. 75–113.

van den Berg, J., Miller, S., Duckworth, D., and Hu, H., (2010). Superhuman performance of surgical tasks by robots using iterative learning from human-guided demonstrations. *Proceedings of 2010 IEEE International Conference on Robotics and Automation*, Anchorage, USA, pp. 2074–2081.

Biggs, G., and MacDonald, B., (2003). A survey of robot programming systems. *Proceedings of Australasian Conference on Robotics and Automation*, Brisbane, Australia, pp. 1–10.

Billard, A., and Hayes, G., (1999). DRAMA, a connectionist architecture for control and learning in autonomous robots. *Adaptive Behavior*, vol. 7, no. 1, pp. 35–64.

Billard, A., Eppars, Y., Calinon, S., Schaal, S., and Cheng, G., (2004). ' Discovering optimal imitation strategies. *Robotics and Autonomous Systems*, vol. **47**, no. 2–3, pp. 69–77.

Billard, A., Calinon, S., Dillmann, R., and Schaal, S., (2008). Robot programming by demonstration. *Handbook of Robotics*. (Eds.) Siciliano, B., and Oussama, K., Berlin, Germany: Springer-Verlag, ch. 59.

Bischoff, R., Kazi, A., and Seyfarth, M., (2002). The MORPHA style guide for icon-based programming. *Proceedings of 11th International Workshop on Robot and Human Interactive Communication*, Augsburg, Germany, pp. 482–487.

Calinon, S., (2009). *Robot Programming by Demonstration: A Probabilistic Approach*. Boca Raton, USA: EPFL/CRC Press.

Calinon, S., and Billard, A., (2004). Stochastic gesture production and recognition model for a humanoid robot. *Proceedings of IEEE/RSJ International Conference on Intelligent Robots and Systems*, Sendai, Japan, pp. 2769–2774.

Calinon, S., and Billard, A., (2006). Teaching a humanoid robot to recognize and reproduce social cues. *Proceedings of IEEE International Symposium on Robot and Human Interactive Communication*, Hatfield, UK, pp. 346–351.

Calinon, S., and Billard, A.G., (2007a). What is the teacher's role in robot programming by demonstration? Toward benchmark for improved learning. *Journal of Interaction Studies*, vol. **8**, no. 3, pp. 441–464.

Calinon, S., and Billard, A., (2007b). Learning of gestures by imitation in a humanoid robot. *Imitation and Social Learning in Robots, Humans and Animals: Social and Communicative Dimension*. (Eds.) Dautenhahn, K., and Nehaniv, C.L., Cambridge, USA: Cambridge University Press, pp. 153–178.

Calinon, S., and Billard, A., (2008). A probabilistic programming by demonstration framework handling constraints in joint space and task space. *Proceedings of 2008 IEEE/RSJ International Conference on Intelligent Robots and Systems*, Nice, France, pp. 367–372.

Calinon, S., Guenter, F., and Billard, A., (2005). Goal-directed imitation in a humanoid robot. *Proceedings of IEEE International Conference on Robotics and Automation*, Barcelona, Spain, pp. 300–305.

Calinon, S., Guenter, F., and Billard, A., (2007). On learning, representing and generalizing a task in a humanoid robot. *IEEE Transaction on Systems, Man and Cybernetic—Part B*, vol. **37**, no. 2, pp. 286–298.

Chan, S., and Liaw, H., (1996). Generalized impedance control of robot for assembly tasks requiring compliant manipulation. *IEEE Transactions on Industrial Electronics*, vol. **43**, no. 4, pp. 453–461.

Chernova, S., and Veloso, M., (2008a). Multi-thresholded approach to demonstration selection for interactive robot learning. *Proceedings of ACME/IEEE International Conference on Human–Robot Interaction*, Amsterdam, the Netherlands, pp. 225–232.

Chernova, S., and Veloso, M., (2008b). Teaching multi-robot coordination using demonstration and state sharing. *Proceedings of International Conference on Autonomous Agents*, Estoril, Portugal, pp. 1183–1186.

Dai, W., and Kampker, M., (2000). User oriented integration of sensor operations in an offline programming system for welding robots. *Proceedings of IEEE International Conference on Robotics and Automation*, San Francisco, USA, pp. 1563–1567.

Dautenhahn, K., and Nehaniv, C.L., (2002). *Imitation in Animals and Artifacts*. Cambridge, USA: MIT Press.

Dillmann, R., (2004). Teaching and learning of robot tasks via observation of human performance. *Robotics and Autonomous Systems*, vol. **47**, no. 2–3, pp. 109–116.

Dillmann, R., Kaiser, M., and Ude, A., (1995). Acquisition of elementary robot skills from human demonstration. *Proceedings of International Symposium on Intelligent Robotic Systems*, Pisa, Italy, pp. 1–38.

Dufay, B., and Latombe, J.-C., (1984). An approach to automatic robot programming based on inductive learning. *International Journal of Robotics Research*, vol. **3**, no. 3, pp. 3–20.

Ehrenmann, M., Zollner, R., Knoop, S., and Dillmann, R., (2001). Sensor fusion approaches for observation of user actions in programming by demonstration. *Proceedings of International Conference on Multisensor Fusion and Integration for Intelligent Systems*, Seoul, Korea, pp. 227–232.

Ehrenmann, M., Zollner, R., Rogalla, O., and Dillmann, R., (2002). Programming service tasks in household environments by human demonstration. *Proceedings of IEEE International Workshop on Robot and Human Interactive Communication*, Berlin, Germany, pp. 460–467.

Ekvall, S., and Kragic, D., (2008). Robot learning from demonstrations: a task-level planning approach. *International Journal of Advanced Robotic Systems*, vol. **5**, no. 3, pp. 223–234.

Ekvall, S., Aarno, D., and Kragic, D., (2006). Task learning using graphical programming and human demonstrations. *IEEE International Symposium on Robot and Human Interactive Communication*, Hatfield, United Kingdom, pp. 398–403.

Fang, H.C., Ong, S.K., and Nee, A.Y.C., (2012). Interactive robot trajectory planning and simulation using Augmented Reality. *Robotics and Computer Integrated Manufacturing*, vol. **28**, no. 2, pp. 227–238.

Field, M., Stirling, D., Pan, Z., and Naghdy, F., (2016). Learning trajectories for robot programming by demonstration using a coordinated mixture of factor analysis. *IEEE Transactions on Cybernetics*, vol. **46**, no. 3, pp. 706–717.

Fischer, K., Kirstein, F., Jensen, L.C., Kruger, N., Kuklinski, K., Wieschen, M.V., *et al.*, (2016). A comparison of types of robot control for programming by demonstration. *Proceedings of 11th ACM/IEEE International Conference on Human–Robot Interaction*, Christchurch, New Zealand, pp. 213–220.

Freese, M., Singh, S., Ozaki, F., and Matsuhira, N., (2010). Virtual robot experimentation platform v-rep: a versatile 3d robot simulator. *Proceedings of Second International Conference on Simulation, Modeling, and Programming for Autonomous Robots*, Berlin, Germany, pp. 51–62.

Friedrich, H., Dillmann, R., and Rogalla, O., (1998). Interactive robot programming based on human demonstration and advice. *International Workshop on Sensor Based Intelligent Robots*, Dagstuhl Castle, Germany, pp. 96–119.

Gams, A., and Ude, A., (2009). Generalization of example movements with dynamic systems. *Proceedings of IEEE/RAS International Conference on Humanoid Robots*, Paris, France, pp. 28–33.

Grebenstein, M., Albu-Schaffer, A., Bahls, T., Chalon, M., Eiberger, O., Friedl, W., *et al.*, (2011). The DLR hand arm system. *Proceedings of 2011 IEEE International Conference on Robotics and Automation*, Shanghai, China, pp. 3175–3182.

Gribovskaya, E., Khansari-Zadeh, S.M., and Billard, A., (2010). Learning nonlinear multivariate dynamics of motion in robotics manipulators. *International Journal of Robotics Research*, vol. **30**, no. 1, pp. 80–117.

Hopler, R., and Otter, M., (2001). A versatile C++ toolbox for model based, real time control systems of robotic manipulators. *Proceedings of IEEE/RSJ International Conference on Intelligent Robots and Systems*, Maui, USA, pp. 2208–2214.

Ijspeert, A.J., Nakanishi, J., and Schaal, S., (2002a). Movement imitation with nonlinear dynamical systems in humanoid robots. *Proceedings of 2002 IEEE International Conference on Robotics and Automation*, Washington, USA, pp. 1398–1403.

Ijspeert, A.J., Nakanishi, J., and Schaal, S., (2002b). Learning rhythmic movements by demonstration using nonlinear oscillators. *Proceedings of*

IEEE/RSJ International Conference on Intelligent Robots and Systems, Lausanne, Switzerland, pp. 958–963.

Ijspeert, A.J., Nakanishi, J., and Schaal, S., (2003). Learning attractor landscapes for learning motor primitives. *Advances in Neural Information Processing Systems 15*. (Eds.) Becker, S., Thrun, S., and Obermayer, K., Cambridge, USA: MIT Press, pp. 1547–1554.

Ikeuchi, K., and Suehiro, T., (1993). Toward an assembly plan from observation. Part I: task recognition with polyhedral objects. *IEEE Transactions on Robotics and Automation*, vol. **10**, no. 3, pp. 368–285.

Inamura, T., Tanie, H., and Nakamura, Y., (2003). Keyframe compression and decompression for time series data based on the continuous hidden Markov model. *Proceedings of IEEE/RSJ International Conference on Intelligent Robots and Systems*, Las Vegas, USA, pp. 1487–1492.

Jafari, A., Tsagarakis, N., and Caldwell, D., (2011). Awas-ii: A new actuator with adjustable stiffness based on the novel principle of adaptable pivot point and variable lever ratio. *Proceedings of 2011 IEEE International Conference on Robotics and Automation*, Shanghai, China, pp. 4638–4643.

Kanayama, Y.T., and Wu, C.T., (2000). It's time to make mobile robots programmable. *Proceedings of IEEE International Conference on Robotics and Automation*, San Francisco, USA, pp. 329–334.

Kim, J.-J., Park, S.-Y., and Lee, J.-J., (2015). Adaptability improvement of learning from demonstration with sequential quadratic programming for motion planning. *Proceedings of IEEE International Conference on Advanced Intelligent Mechatronics (AIM)*, Busan, Korea, pp. 1032–1037.

Kjellstrom, H., Romero, J., and Kragic, D., (2008). Visual recognition of grasps for human-to-robot mapping. *Proceedings of IEEE/RSJ International Conference on Intelligent Robots and Systems*, Nice, France, pp. 3192–3199.

Koenig, N., and Howard, A., (2004). Design and use paradigms for gazebo, an open-source multi-robot simulator. *Proceedings of 2004 IEEE/RSJ International Conference on Intelligent Robots and Systems*, Sendai, Japan, pp. 2149–2154.

Konidaris, G., Barto, A., Grupen, R., and Kuindersma, S. (2012). Robot learning from demonstration by constructing skill trees. *International Journal of Robotics Research*, vol. 3, no. 1, pp. 360–375.

Kronander, K., and Billard, A., (2014). Learning compliant manipulation through kinesthetic and tactile human–robot interaction. *IEEE Transactions on Haptics*, vol. 7, no. 3, pp. 367–380.

Kruger, V., Herzog, D.L., Baby, S., Ude, A., and Kragic, D., (2010). Learning actions from observations: primitive-based modeling and grammar. *Robotics and Automation Magazine*, vol. **17**, no. 2, pp. 30–43.

Lauria, S., Bugmann, G., Kyriacou, T., and Klein, E., (2002). Mobile robot programming using natural language. *Robotics and Autonomous Systems*, vol. **38**, no. 3, pp. 171–181.

Lemme, A., Freire, A., Barreto, G., and Steil, J., (2013). Kinesthetic teaching of visuomotor coordination for pointing by the humanoid robot icub. *Neurocomputing*, vol. **112**, no. 3, pp. 179–188.

Liu, S., and Asada, H., (1993). Teaching and learning of deburring robots using neural networks. *Proceedings of IEEE International Conference on Robotics and Automation*, Taipei, Taiwan, pp. 339–345.

Liu, P., Glas, D.F., Ishiguro, H., and Hagita, N., (2014). How to train your robot—Teaching service robots to reproduce human social behavior. *Proceedings of 23rd IEEE International Symposium on Robot and Human Interactive Communication*, Edinburgh, UK, pp. 961–968.

Lozano-Pérez, T, (1983). Robot programming. *Proceedings of the IEEE*, vol. **71**, no. 7, pp. 821–841.

Luger, G.F., (2008). *Artificial Intelligence: Structures and Strategies for Complex Problem Solving* (sixth edition). Boston, USA: Pearson Education Inc., pp. 442–449.

Martinez, D., and Kragic, D., (2008). Modeling and recognition of actions through motor primitives. *Proceedings of IEEE International Conference on Robotics and Automation*, Pasadena, USA, pp. 1704–1709.

McGuire, P., Fritsch, J., Steil, J., Rothling, F., Fink, G., Wachsmuth, S., *et al.*, (2002). Multi-modal human–machine communication for instructing robot grasping tasks. *Proceedings of IEEE/RSJ International Conference on Intelligent Robots and Systems*, Lausanne, Switzerland, pp. 1082–1088.

Munch, S., Kreuziger, J., Kaiser, M., and Dillmann, R., (1994). Robot programming by demonstration (RPD)—using machine learning and user interaction methods for the development of easy and comfortable robot programming systems. *Proceedings of 24th International Symposium on Industrial Robots*, Hannover, Germany, pp. 685–693.

Mylonas, G.P., Giataganas, P., Chaudery, M., Vitiello, V., Darzi, A., and Yang, G.Z., (2013). Autonomous eFAST ultrasound scanning by a robotic manipulator using learning from demonstrations. *Proceedings of 2013 IEEE/RSJ International Conference on Intelligent Robots and Systems*, Tokyo, Japan, pp. 3251–3256.

Nakamura, Y., and Hanafusa, H., (1986). Inverse kinematic solutions with singularity robustness for robot manipulator control. *ASME Journal of Dynamic Systems, Measurement, and Control*, vol. **108**, no. 3, pp. 163–171.

Nathanael, D., Mosialos, S., Vosniakos, G.C., (2016). Development and evaluation of a virtual training environment for on-line robot programming. *International Journal of Industrial Ergonomics*, vol. **53**, pp. 274–283.

Nehaniv, C.L., and Dautenhahn, K., (2001). Like me?—Measures of correspondence and imitation. *International Journal on Cybernetics and Systems*, vol. **32**, pp. 11–52.

Ng, C.W.X., Chan, K.H.K., Teo, W.K., and I-Ming, C., (2014). A method for capturing the tacit knowledge in the surface finishing skills by demonstration for programming a robot. *Proceedings of 2014 IEEE International Conference on Robotics and Automation*, Hong Kong, China, pp. 1374–1379.

Nicolescu, M.N., and Mataric, M.J., (2001). Experience-based representation construction: learning from human and robot teachers. *Proceedings of IEEE/RSJ International Conference on Intelligent Robots and Systems*, Maui, USA, pp. 740–745.

Ogawara, K., Takamatsu, J., Kimura, H., and Ikeuchi, K., (2002a). Generation of a task model by integrating multiple observations of human demonstrations. *Proceedings of IEEE International Conference on Robotics and Automation*, Washington, USA, pp. 1545–1550.

Ogawara, K., Takamatsu, J., Kimura, H., and Ikeuchi, K., (2002b). Modeling manipulation interactions by hidden Markov models. *Proceedings of IEEE/ RSJ International Conference on Intelligent Robots and Systems*, Lausanne, Switzerland, pp. 1096–1101.

Ogawara, K., Takamatsu, J., Kimura, H., and Ikeuchi, K. (2003). Extraction of essential interactions through multiple observations of human demonstrations. *IEEE Transactions on Industrial Electronics*, vol. **50**, no. 4, pp. 667–675.

Ogino, M., Toichi, H., Yoshikawa, Y., and Asada, M., (2006). Interaction rule learning with a human partner based on an imitation faculty with a simple visuo-motor mapping. *Robotics and Autonomous Systems*, vol. **56**, no. 5, pp. 414–418.

Payandeh, S., and Stanisic, Z., (2002). On application of virtual fixtures as an aid for telemanipulation and training. *Proceedings of 10th Symposium on Haptic Interfaces for Virtual Environment and Teleoperator Systems*, Orlando, USA, pp. 18–23.

Peternel, L., Petric, T., Oztop, E., and Babič, J., (2014). Teaching robots to cooperate with humans in dynamic manipulation tasks based on multi-modal human in-the-loop approach. *Autonomous Robots*, vol. **36**, no. 1–2, pp. 123–136.

Pomplun, M., and Mataric, M.J., (2000). Evaluation metrics and results of human arm movement imitation. *Proceedings of First IEEE-RAS International Conference on Humanoid Robotics*, Cambridge, USA, pp. 1–8.

Pook, P.K., and Ballard, D.H., (1993). Recognizing teleoperated manipulations. *Proceedings of IEEE International Conference on Robotics and Automation*, Atlanta, USA, pp. 578–585.

Rooks, B.W., (1997). Off-line programming: a success for the automotive industry. *International Journal of Industrial Robot*, vol. **24**, no. 1, pp. 30–34.

Rosenstein, M.T., and Barto, A.G. (2004). Supervised actor-critic reinforcement learning. *Learning and Approximate Dynamic Programming: Scaling Up to the Real World*. (Eds.) Si, J., Barto, A., Powell, W., and Wunsch, D., New York, USA: John Wiley & Sons, Inc., ch. 14.

Rozo, L., Calinon, S., Caldwell, D.C., Jiménez, P., and Torras, C., (2016). Learning physical collaborative robot behaviors from human demonstrations. *IEEE Transactions on Robotics*, vol. **32**, no. 3, pp. 513–527.

Saunders, J., Nehaniv, C.L., and Dautenhahn, K., (2006). Teaching robots by moulding behavior and scaffolding the environment. *Proceedings of ACM/ IEEE International Conference on Human–Robot Interaction*, Salt Lake City, USA, pp. 118–125.

Scassellati, B., (1999). Imitation and mechanisms of joint attention: a developmental structure for building social skills on a humanoid robot. *Computation for Metaphors, Analogy and Agents*. (Ed.) Nehaniv, C., vol. **1562**, Springer Lecture Notes in Artificial Intelligence, Berlin, Germany: Springer, pp. 176–195.

Schaal, S., and Atkeson, C.G., (1998). Constructive incremental learning from only local information. *Neural Computation*, vol. **10**, no. 8, pp. 2047–2084.

Schneider, M., and Ertel, W., (2010). Robot learning by demonstration with local Gaussian process regression. *Proceedings of IEEE/RAS International Conference on Intelligent Robots and Systems*, Taipei, Taiwan, pp. 255–260.

Seidel, D., Emmerich, C., and Steil, J.J., (2014). Model-free path planning for redundant robots using sparse data from kinesthetic teaching. *Proceedings of 2014 IEEE/RSJ International Conference on Intelligent Robots and Systems*, Chicago, USA, pp. 4381–4388.

Shimizu, M., Endo, Y., Onda, H., Yoon, W.-K., and Torii, T., (2013). Improving task skill transfer method by acquiring impedance parameters from human demonstrations. *Proceedings of 2013 IEEE International Conference on Mechatronics and Automation*, Takamatsu, Japan, pp. 1033–1038.

Steinfeld, A., Fong, T., Kaber, D., Lewis, M., Scholtz, J., Schultz, A., *et al.*, (2006). Common metrics for human–robot interaction. *Proceedings of ACM/IEEE International Conference on Human–Robot Interaction*, Salt Lake City, USA, pp. 33–40.

Takahashi, T., and Sakai, T. (1991). Teaching robot's movement in virtual reality. *Proceedings of International Conference on Robotics and Automation*, Osaka, Japan, pp. 1538–1587.

Thamma, M., Huang, L.H., Lou, S.-J., and Diez, C.R., (2004). Controlling robot through internet using Java. *Journal of Industrial Technology*, vol. **20**, no. 3, pp. 1–6.

Thomaz, A.L., and Breazeal, C., (2006). Reinforcement learning with human teachers: evidence of feedback and guidance with implications for learning performance. *Proceedings of 21st National Conference on Artificial Intelligence*, Boston, USA, pp. 1000–1005.

Tso, S.K., and Liu, K.P., (1997). Demonstrated trajectory selection by hidden Markov model. *Proceedings of International Conference on Robotics and Automation*, Albuquerque, USA, pp. 2713–2718.

Vijayakumar, S., and Schaal, S., (2000). Locally weighted projection regression: an $O(n)$ algorithm for incremental real time learning in high dimensional space. *Proceedings of International Conference on Machine Learning*, San Francisco, USA, pp. 1079–1086.

Voyles, R., and Khosla, P., (1999). Gesture-based programming: a preliminary demonstration. *Proceedings of IEEE International Conference on Robotics and Automation*, Detroit, USA, pp. 708–713.

Wachter, M., Schulz, S., Asfour, T., Aksoy, E., Worgotter, F., and Dillmann, R., (2013). Action sequence reproduction based on automatic segmentation and object-action complexes. *Proceedings of IEEE-RAS International Conference on Humanoid Robots*, Atlanta, USA, pp. 189–195.

Wampler, C.W., (1986). Manipulator inverse kinematic solutions based on damped least-squares solutions. *IEEE Transactions on Systems, Man, and Cybernetics*, vol. **16**, no. 1, pp. 93–101.

Webel, S., Bockholt, U., Engelke, T., Gavish, N., Olbrich, M., and Preusche, C., (2013). An augmented reality training platform for assembly and maintenance tasks. *Robotics and Autonomous Systems*, vol. **61**, pp. 398–403.

Whitney, D.E., (1969). Resolved motion rate control of manipulators and human prostheses. *IEEE Transactions on Man-Machine Systems*, vol. **10**, no. 2, pp. 47–53.

Whitney, D.E., (1972). The mathematics of coordinated control of prosthetic arms and manipulators. *ASME Journal of Dynamic Systems, Measurement and Control*, vol. **94**, no. 4, pp. 303–309.

Wu, X., and Kofman, J., (2008). Human-inspired robot task learning from human teaching. *Proceedings of IEEE International Conference on Robotics and Automation*, Pasadena, USA, pp. 3334–3339.

Yang, J., Xu, Y., and Chen, C.S., (1994). Hidden Markov model approach to skill learning and its application to telerobotics. *IEEE Transactions on Robotics and Automation*, vol. **10**, no. 5, pp. 621–631.

Yang, J., Xu, Y., and Chen, C.S., (1997). Human action learning via hidden Markov model. *IEEE Transactions on Systems, Man, and Cybernetics—Part A*, vol. **27**, no. 1, pp. 34–44.

Yang, Y., Lin, L.L., Song, Y.T., Nemec, B., Ude, A., Buch, A.G., *et al.*, (2014). Fast programming of peg-in hole actions by human demonstrations. *Proceedings of 2014 International Conference on Mechatronics and Control*, Jinzhou, China, pp. 990–995.

2

Task Perception

For capturing the relevant motions during the demonstrations, two common perception interfaces are presented in the book: optical tracking systems and vision cameras. The acquired data from optical trackers are temporal sequences of positions and orientations of the scene objects. Vision cameras acquire sequence of images of the scene, thus the task representation requires reduction of the dimensionality of the acquired data via extraction of relevant image features. Trajectories of the relevant objects during the demonstrations are commonly calculated by employing a subset of the extracted image features.

2.1 Optical Tracking Systems

An optical tracking system, Optotrak Certus® from NDI (Waterloo, Canada) (Optotrak Certus, 2012), is shown in Figure 2.1a. The optical tracking system employs a set of infrared *optical markers* for capturing the demonstrated motions (Figure 2.1b). The markers are attached onto strategic locations of the target objects, tools, or the demonstrator's body. Based on the markers' locations with respect to a fixed reference frame, poses of predefined rigid bodies are inferred over a set of discrete time instants. For the considered tasks here, the Optotrak system is set to acquire the positions of the optical markers at predefined time periods of 10 milliseconds. According to the manufacturer's data sheets, the accuracy of the Optotrak system, at a distance of 2 meters from the position sensors, is characterized by the root-mean-square (RMS) errors of the horizontal and vertical measurements in the range of 0.1 millimeters, whereas for the depth coordinate, the RMS error is approximately 0.15 millimeters. The resolution of the measurements is 0.01 millimeter. The full focus

Robot Learning by Visual Observation, First Edition. Aleksandar Vakanski and Farrokh Janabi-Sharifi.
© 2017 John Wiley & Sons, Inc. Published 2017 by John Wiley & Sons, Inc.

(a)

(b)

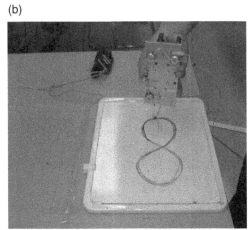

Figure 2.1 (a) Position camera sensors of the optical tracking system Optotrak Certus; (b) optical markers attached on a tool are tracked during a demonstration of a "figure 8" motion.

measurement volume of the system is: 4.2 meters in width × 3 meters in height × 7 meters in length.

The perceptual streams consist of ordered sets of position and orientation data for each object of interest, that is, $\{X_m = (\mathbf{P}_{m,l}(t_k), \boldsymbol{\varphi}_{m,l}(t_k))_{k=1}^{T_m}\}_{m=1}^{M}$, for $l = 1, 2, ..., L$, where the notation symbol l is used for indexing the tracked objects. The position vectors $\mathbf{P}_{m,l}(t_k) \in \mathbb{R}^3$ pertain to the Cartesian x, y, and z coordinates of the rigid bodies, and the orientation vectors $\boldsymbol{\varphi}_{m,l}(t_k) \in \mathbb{R}^3$ consist of the Euler roll–pitch–yaw angles, acquired at the time instant t_k. Note that in the text that follows, rotation matrices corresponding to sets of Euler angles will also be used for representation of the orientation whenever required.

2.2 Vision Cameras

A Firefly® MV CMOS camera from Point Grey (Richmond, Canada) (Firefly MV, 2012) has also been used for the observation of demonstrations in the book. The size of the camera's image sensor is 1/3 inches, with a resolution of 640×480 pixels. The camera enables image acquisition with a frame rate of up to 60 Hz. A lens with 8 millimeters focal distance from Edmund Optics (Barrington, USA) has been used with the camera.

The task perception with vision cameras is performed via acquiring visual information from the environment in the form of sequences of images. For a vision camera with a resolution of $u_r \times v_r$ pixels, the measurement data vectors at each time instant are high dimensional, that is, $\mathbf{x}_m(t_k) \in \mathbb{R}^{u_r \times v_r}$. Representing the acquired data in a low-dimensional space for learning purposes requires processing the images and extraction of relevant *image features*. The image features are often selected as projections of distinctive three-dimensional (3D) attributes from the scene onto the image plane. Examples are corners and edges of the manipulated objects or tools, artificial markers placed on the demonstrator's hand, etc. The measurable properties of the image features are referred to as *image features parameters*.

Coordinates of points in the image plane are employed as image features here, although other types, such as distances between points, line lengths, and image moments, have also been used in the literature (Hutchinson *et al.*, 1996). The total number of image features extracted from the grabbed images is denoted with N, while the total number of demonstrations is M. The observed horizontal and vertical pixel coordinates (u and v, respectively) of the feature point n at time t_k for the demonstration m are denoted by $\mathbf{u}_m^{(n)}(t_k) = \left[u_m^{(n)}(t_k) \ \ v_m^{(n)}(t_k) \right]^T \in \mathbb{R}^2$, for $n = 1, 2, ..., N$, $m = 1, 2, ..., M$.

Furthermore, under an assumption that a calibrated camera is used, the extracted coordinates of the feature points can be transformed from pixel coordinates into a pair of horizontal and vertical spatial image plane coordinates (Janabi-Sharifi, 2002; Chaumette and Hutchinson, 2006), i.e.,

$$
\begin{cases}
r_m^{u,(n)}(t_k) = \dfrac{u_m^{(n)}(t_k) - u_0}{f_c k_u} \\[4mm]
r_m^{v,(n)}(t_k) = \dfrac{v_m^{(n)}(t_k) - v_0}{f_c k_u}
\end{cases}, \tag{2.1}
$$

where u_0 and v_0 are the horizontal and vertical pixel coordinates of the principal point of the image plane, f_c denotes the focal length of the camera, and k_u and k_v are the horizontal and vertical number of pixels per unit length. The pairs of horizontal and vertical image plane coordinates of the feature point n are denoted by $\mathbf{r}_m^{(n)}(t_k) = \left[r_m^{u,(n)}(t_k) \ \ r_m^{v,(n)}(t_k) \right]^T \in \mathbb{R}^2$. The vector consisting of all observed features at time t_k for the demonstration m forms the *image feature parameters vector*, denoted by

$\xi_m(t_k) = \left[\mathbf{r}_m^{(1)}(t_k)^T \quad \mathbf{r}_m^{(2)}(t_k)^T \quad \ldots \quad \mathbf{r}_m^{(n)}(t_k)^T \right]^T \in \mathbb{R}^{2n}$. The set of image features trajectories from all M demonstrations comprises the demonstration data $\{\xi_m(t_1), \xi_m(t_2), \ldots, \xi_m(t_{T_m})\}_{m=1}^{M}$.

It is further assumed that geometric models of the scene objects are available, providing knowledge about the 3D distances between the features of objects. Then, for a given vector of feature coordinates in the image plane $\xi_{m,l}(t_k)$ of an object with index l, and using the available information for the distances between the corresponding features in the Cartesian space, a homography transformation (Faugeras, 1993) can be employed for extraction of the object's pose with respect to the camera $(\mathbf{P}_{m,l}(t_k), \mathbf{R}_{m,l}(t_k))$. For estimation of the homography matrix, the correspondences of at least four coplanar points or eight noncoplanar points are required. The poses relate to the kinematics of rigid bodies in the 3D space, that is, they belong to the special Euclidean group of index 3, $(\mathbf{P}_{m,l}(t_k), \mathbf{R}_{m,l}(t_k)) \in SE(3)$. As noted earlier, the set of Euler's roll–pitch–yaw angles $\varphi_{m,l}(t_k) \in \mathbb{R}^3$ corresponding to the rotation matrix $\mathbf{R}_{m,l}(t_k)$ will be interchangeably used throughout the text for representation of the orientation.

2.3 Summary

The chapter contains the review of two data acquisition systems commonly used for perception of demonstrated motions. Optical tracking systems employ a set of infrared optical markers attached on the object of interest, whose position with respect to cameras is used for calculating the demonstrated trajectories. Vision cameras rely on light intensity images from the environment for motion perception. Image processing algorithms are employed subsequently for dimensionality reduction through image features extraction and pose estimation. While the optical tracking systems provide accurate and reliable means for capturing demonstrated tasks, the vision cameras allow natural and unrestrained motions during the demonstrations which can quickly be performed without the need for attaching markers on the objects in the scene.

References

Chaumette, F., and Hutchinson, S., (2006). Visual servo control—part I: basic approaches. *IEEE Robotics and Automation Magazine*, vol. **13**, no. 4, pp. 82–90.

Faugeras, O., (1993). *Three-Dimensional Computer Vision: A Geometric Viewpoint.* Cambridge, USA: MIT Press.

Firefly MV. *CMOS Camera*, Point Grey, (2012). Available from: http://www.ptgrey.com/products/fireflymv/fireflymv_usb_firewire_cmos_camera.asp (accessed on September 7, 2016).

Hutchinson, S., Hager, G., and Corke, P., (1996). A tutorial on visual servo control. *IEEE Transactions on Robotics and Automation*, vol. **12**, no. 5, pp. 651–670.

Janabi-Sharifi, F., (2002) Visual servoing: Theory and applications. *Opto-Mechatronics Systems Handbook.* (Ed.) Cho, H., Boca Raton, USA: CRC Press, ch. 15.

Optotrak Certus. *Motion Capture System*, Northern Digital Inc. (2012). Available from: http://www.ndigital.com/lifesciences/certus-motioncapturesystem.php (accessed on September 7, 2016).

3

Task Representation

This chapter reviews approaches for task representation in robot programming by demonstration (PbD). Based on the level of task abstraction, the methods are categorized into high-level task representation at the symbolic level of abstraction and low-level task representation at the trajectory level of abstraction. Techniques for data preprocessing related to trajectories scaling and aligning are also discussed in the chapter.

The PbD framework aims at learning from multiple demonstrations of a skill performed under similar conditions. For a set of M demonstrations, the perception data are denoted by $\{X_m = (\mathbf{x}_m(t_1), \mathbf{x}_m(t_2), ..., \mathbf{x}_m(t_{T_m}))\}_{m=1}^{M}$, where m is used for indexing the demonstrations, t is used for indexing the measurements within each demonstration, and T_m denotes the total number of measurements of the demonstration sequence X_m. Each measurement represents a D-dimensional vector $\mathbf{x}_m(t_k) = \left[x_m^{(1)}(t_k) \ \ x_m^{(2)}(t_k) \ \ ... \ \ x_m^{(D)}(t_k) \right]^T$. The form of the measurements depends on the data acquisition system(s) employed for perception of the demonstrations, and it can encompass the following:

a) Cartesian poses (positions and orientations) and/or velocities/accelerations of manipulated objects, tools, or the demonstrator's hand.
b) Joint angles positions and/or velocities/accelerations of the demonstrator's arm.
c) Forces exerted on the environment by the demonstrated actions.
d) Sequence of images of the scene in the case of vision-based perception.

The data structure for the task perception with optical sensors and vision cameras is presented in Sections 2.1 and 2.2.

The objective of PbD is to map the demonstrated space into a sequence of low-level command signals for a robot learner to reproduce the

Robot Learning by Visual Observation, First Edition. Aleksandar Vakanski and Farrokh Janabi-Sharifi.

demonstrated task. As outlined in Chapter 2, the mapping is often highly nonlinear, and therefore, it represents a challenging problem. It usually consists of several steps. The task modeling and task analysis phases encode the recorded data into compact and flexible representation of demonstrated motions and extract the relevant task features for achieving the required robot performance. In the task planning step, a trajectory for task reproduction is derived by generalization over the dataset of demonstrated task examples. The generalized trajectory for task reproduction is denoted by $X_{\text{gen}} = \left(\mathbf{x}_{\text{gen}}(t_1), \mathbf{x}_{\text{gen}}(t_2), ..., \mathbf{x}_{\text{gen}}(t_{T_{\text{gen}}})\right)$, where T_{gen} is the number of time measurements of the generalized trajectory. Afterward, the generalized trajectory is transferred to robot commands and loaded on the robot platform for the task execution in the real environment.

3.1 Level of Abstraction

Representation of observed tasks in robot PbD is often categorized based on the level of task abstraction into high-level and low-level representation (Schaal *et al.*, 2003). As elaborated in Section 1.4.2, high-level task representation refers to encoding tasks as a hierarchical sequence of high-level behaviors (Dillmann, 2004; Saunders *et al.*, 2006). This representation category is also known as symbolic task representation. In general, the elementary behaviors are predefined, and the observed tasks are initially segmented into sequences of behavior. The task representation then involves defining rules and conditions related to the state of the required world for each elementary action to occur, as well as defining rules and required conditions for each elementary action to end and the successor behavior to begin. Among the principal drawbacks for such task representation approach is the limitation of relying on a library of predefined elementary behaviors in encoding tasks, and the limited applicability for representing tasks requiring continuous high accuracy along a three-dimensional (3D) path or a velocity profile.

Low-level task representation is also referred to as representation at a trajectory level, where the task demonstrations are encoded as continuous 3D trajectories in the Cartesian space or as continuous angle trajectories in the joint space (Section 1.4.2.2). The PbD approaches presented in the book employ this type of task abstraction, due to the ability to represent arbitrary motions, as well as to define the spatial constraints of the demonstrations (Aleotti and Caselli, 2005; Calinon, 2009). Additionally, low-level task representation enables to specify the velocities and accelerations at different phases of the demonstrated task, and allows encoding tasks with high-accuracy demand.

3.2   Probabilistic Learning

Statistical methods have been widely used in robotics for representing the uncertain information about the state of the environment. On one hand, processing of the robot's sensory data involves handling the uncertainties that originate from sensors noise, limitations of the sensors, and unpredictable changes in the environment. On the other hand, processing the control actions involves tackling the uncertainties regarding the way the decisions are made to achieve the desired level of motor performance. The statistical frameworks represent the uncertainties of robot's perception and action via probability distributions, instead of using a single best guess about the state of the world. As a result, the use of statistical models has contributed to improved robustness and performance in many robotic applications, such as localization and navigation of mobile robots, planning, and map generation from sensory data (Thrun, 2000).

Regarding robot learning from observation of human demonstrations, the theory of statistical modeling has been exploited for representing the uncertainties of the acquired perceptual data (Calinon, 2009). Indeed, one must bear in mind that a fundamental property of the human movements is their random nature. Humans are not capable of drawing perfect straight lines or repeating identical movements, which is assumed is due to the inherent stochastic nature in the neural information processing of the required actions (Clamann, 1969). The statistical algorithms provide a form to encapsulate the random variations in the observed demonstrations by deriving the probability distributions of the outcomes from several repeated measurements. Consequently, a model of a human skill is built from several examples of the same skill demonstrated in a similar fashion and under similar conditions. The underlying variability across the repeated demonstrations is utilized to probabilistically represent the different components of the task, and subsequently, to retrieve a generalized version of the demonstrated trajectories.

Within the published literature, hidden Markov model (HMM) has been established as one of the most popular statistical methods for modeling of human motions. Other approaches that have been used for probabilistic skill encoding include the following: Gaussian mixture model (GMM) (Calinon, 2009), support vector machines (Zollner *et al.*, 2002; Martinez and Kragic, 2008), and Bayesian belief networks (Coates *et al.*, 2008; Grimes and Rao, 2008).

3.3   Data Scaling and Aligning

Generalization from multiple task examples imposes the need to address the problem of temporal variability in the demonstrated data. Often, it is required to initially scale the demonstrated data to sequences with equal length before proceeding with the data analysis. The approaches of linear scaling and dynamic time warping (DTW) scaling have been employed in many works for this purpose (Pomplun and Mataric, 2000; Gribovskaya and Billard, 2008; Ijspeert *et al.*, 2012).

3.3.1 Linear Scaling

Linear scaling refers to the change in the number of measurements of a sequence through interpolation between the sequence data. For a set of collected demonstration trajectories with a different number of time frames, scaling to a new set of sequences with equal number of time frames is achieved by performing linear scaling of the time vectors for each individual trajectory. This procedure is equivalent to extending or shortening the sequences to the required number of time measurements. Among different types of interpolation techniques for time series data, polynomial interpolation is the most often used. More specifically, low-order polynomials are locally fitted to create a smooth and continuous function, referred to as a spline.

For illustration, Figure 3.1a shows two sample trajectories with different number of measurements, whereas Figure 3.1b displays their counterparts after the linear time scaling. Accordingly, the length of the test trajectory is scaled to the number of measurements of the reference trajectory. For sequences that differ significantly in length, this method might not be very efficient, since the temporal variations in the demonstrations can result in spatial misalignments across the set.

3.3.2 Dynamic Time Warping (DTW)

DTW is an algorithm for comparing and aligning time series, where finding an optimal alignment of sequences is based on a nonlinear time warping by using a dynamic programming technique. The DTW alignment for the two sequences from Figure 3.1a is illustrated in Figure 3.1c. It can thus be concluded that the main advantage of the DTW scaling over the linear time scaling is the efficient alignment of the signals for handling the spatial distortions.

The DTW sequence alignment is based on forming a matrix of distances between two time series, and finding an optimal path through the matrix that locally minimizes the distance between the sequences. For a given reference sequence $X_1 = (\mathbf{x}_1(t_1), \mathbf{x}_2(t_2), ..., \mathbf{x}_1(t_{T_\chi}))$ of length T_χ and a test sequence $X_2 = (\mathbf{x}_2(t_1), \mathbf{x}_2(t_2), ..., \mathbf{x}_2(t_{T_\gamma}))$ of length T_γ, the distance matrix $\mathbf{H}$ is formed as follows:

$$\mathbf{H}(\chi, \gamma) = \|\mathbf{x}_1(t_\chi) - \mathbf{x}_2(t_\gamma)\|, \text{ for } \chi = 1, 2, ..., T_\chi, \quad \gamma = 1, 2, ..., T_\gamma.$$
$$(3.1)$$

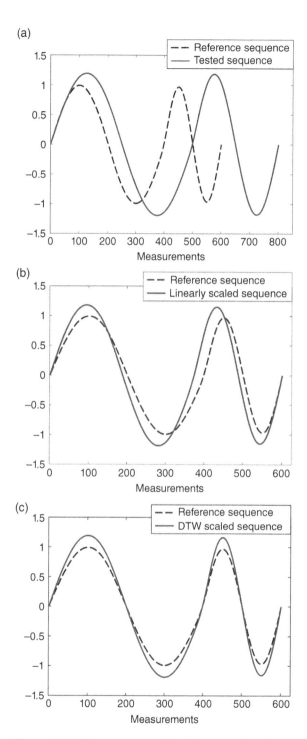

Figure 3.1 (a) Two sequences with different number of measurements: a reference sequence of 600 measurement data points and a test sequence of 800 measurement data points; (b) the test sequence is linearly scaled to the same number of measurements as the reference sequence; and (c) the test sequence is aligned with the reference sequence using DTW.

In (3.1), the notation $\|\cdot\|$ is used for denoting the Euclidean l_2-norm. Note that other norms can also be used as a distance measure, such as l_1-norm (Holt *et al.*, 2007). The optimal alignment path $G = \{g_w(\chi, \gamma)\}$ is a function of the elements of the matrix $\mathbf{H}$ with $\max(T_\chi, T_\gamma) \leq w < T_\chi + T_\gamma - 1$. The path is calculated by minimizing the cumulative sum of the distances $\mathbf{H}(\chi, \gamma)$ and the minimum distance between the test and reference sequences of the neighboring cells, i.e.,

$$g(\chi, \gamma) = \mathbf{H}(\chi, \gamma) + \min\{g(\chi - 1, \gamma), g(\chi - 1, \gamma - 1), g(\chi, \gamma - 1)\}. \quad (3.2)$$

The procedure for implementing DTW includes the following steps:

1) Form the distance matrix of dimensions $T_\chi \times T_\gamma$ for given reference and test sequences using (3.1).
2) Initialize the cumulative distance $g(1,1) = 0$.
3) For $\chi = 2, ..., T_\chi$ and $\gamma = 2, ..., T_\gamma$, calculate the cumulative distances $g(\chi, \gamma)$ according to (3.2).
4) Backtrack the path $G = g_w(\chi, \gamma)$ to find the indices of the warped sequence.

The following constraints are enforced for the warping path:

Boundary conditions: $g_1 = (1,1)$ and $g_{end} = (T_\chi, T_\gamma)$.
Continuity conditions: if $g_w = (\chi, \gamma)$, then $g_{w-1} = (\chi', \gamma')$ for $\chi - \chi' \leq 1$ and $\gamma - \gamma' \leq 1$.
Monotonicity condition: if $g_w = (\chi, \gamma)$, then $g_{w-1} = (\chi', \gamma')$ for $\chi \geq \chi'$ and $\gamma \geq \gamma'$.

The boundary conditions define the starting and ending points of the path, whereas the continuity and monotonicity conditions constrain the path to be continuous and monotonically spaced in time, respectively.

The DTW alignment can excessively distort the signals in the case of dissimilar sequences, therefore shape-preserving constraints are often imposed. Sakoe and Chiba (1978) proposed to force the warped time waves to stay within a specified time frame window of their normalized time vector, and to impose a constraint on the slope of the warping path.

1D DTW refers to alignment of time series based on a single dimension of the data. When dealing with multidimensional data, it might result in suboptimal alignment across the rest of the dimensions. Multidimensional DTW approach was proposed by Holt *et al.* (2007), where the alignment of the sequences involves the coordinates from all dimensions of the data. Hence, the distance matrix includes the Euclidean distance among the dimensions of the sequences at each time point, and has the following form:

$$\mathbf{H}(\chi,\gamma) = \sum_{d=1}^{D} \left\| \mathbf{x}_1^{(d)}(t_\chi) - \mathbf{x}_2^{(d)}(t_\gamma) \right\|, \quad \text{for } \chi = 1, 2, ..., T_\chi, \quad \gamma = 1, 2, ..., T_\gamma,$$

(3.3)

where d denotes the dimensionality of the sequences.

3.4 Summary

Chapter 3 begins with the formulation of the problem of learning trajectories in a PbD setting. Further, methods for task representation at a high-level and low-level of abstraction are covered. A brief overview of statistical methods for task representation is presented. Their application in robot learning from multiple demonstrations is intuitive and natural with consideration to the stochastic character of human motions. The chapter concludes with a discussion of two standard approaches employed for tackling the temporal variations of human-demonstrated trajectories.

References

Aleotti, J., and Caselli, S., (2005). Trajectory clustering and stochastic approximation for robot programming by demonstration. *Proceedings of IEEE/RSJ International Conference on Intelligent Robots and Systems*, Edmonton, Canada, pp. 1029–1034.

Calinon, S., (2009). *Robot Programming by Demonstration: A Probabilistic Approach*. Boca Raton, USA: EPFL/CRC Press.

Clamann, H.P., (1969). Statistical analysis of motor unit firing patterns in a human skeletal muscle. *Journal of Biophysics*, vol. **9**, no. 10, pp. 1223–1251.

Coates, A., Abbeel, P., and Ng, A.Y., (2008). Learning for control from multiple demonstrations. *Proceedings of International Conference on Machine Learning*, Helsinki, Finland, pp. 144–151.

Dillmann, R., (2004). Teaching and learning of robot tasks via observation of human performance. *Robotics and Autonomous Systems*, vol. **47**, no. 2–3, pp. 109–116.

Gribovskaya, E., and Billard, A., (2008). Combining dynamical systems control and programming by demonstration for teaching discrete bimanual coordination tasks to a humanoid robot. *Proceedings of ACME/IEEE*

International Conference on Human-Robot Interaction, Amsterdam, the Netherlands, pp. 1–8.

Grimes, D.B., and Rao, R.P.N., (2008). Learning nonparametric policies by imitation. *Proceedings of IEEE/RSJ International Conference on Intelligent Robots and Systems*, Nice, France, pp. 2022–2028.

Holt, G.A.T., Reinders, M.J.T., and Hendriks, E.A., (2007). Multi-dimensional dynamic time warping for gesture recognition. *Proceedings of 13th Annual Conference of the Advanced School for Computing and Imaging*, Heijen, the Netherlands, pp. 1–8.

Ijspeert, A.J., Nakanishi, J., Hoffmann, H., Pastor, P., and Schaal, S., (2012). Dynamical movement primitives: Learning attractor models for motor behaviors. *Neural Computation*, vol. **25**, no. 2, pp. 328–373.

Martinez, D., and Kragic, D., (2008). Modeling and recognition of actions through motor primitives. *Proceedings of IEEE International Conference on Robotics and Automation*, Pasadena, USA, pp. 1704–1709.

Pomplun, M., and Mataric, M.J., (2000). Evaluation metrics and results of human arm movement imitation. *Proceedings of First IEEE-RAS International Conference on Humanoid Robotics*, Cambridge, USA, pp. 1–8.

Sakoe, H., and Chiba, S., (1978). Dynamic programming algorithm optimization for spoken word recognition. *IEEE Transactions on Acoustics, Speech, and Signal Processing*, vol. **26**, no. 1, pp. 43–49.

Saunders, J., Nehaniv, C.L., and Dautenhahn, K., (2006). Teaching robots by moulding behavior and scaffolding the environment. *Proceedings of ACM/ IEEE International Conference on Human-Robot Interaction*, Salt Lake City, USA, pp. 118–125.

Schaal, S., Ijspeert, A., and Billard, A., (2003). Computational approaches to motor learning by imitation. *Philosophical Transactions of the Royal Society of London: Biological Sciences*, vol. **358**, no. 1431, pp. 537–547.

Thrun, S., (2000). Probabilistic algorithms in robotics. *AI Magazine*, vol. **21**, no. 4, pp. 93–109.

Zollner, R., Rogalla, O., Dillmann, R., and Zollner, M., (2002). Understanding users intentions: programming fine manipulation tasks by demonstration. *Proceedings of IEEE/RSJ International Conference on Intelligent Robots and Systems*, Lausanne, Switzerland, pp. 1114–1119.

4

Task Modeling

The core programming-by-demonstration (PbD) methods for task modeling at a trajectory level of task abstraction are presented in this chapter. The emphasis is on the statistical methods for modeling demonstrated expert actions. The theoretical background behind the following approaches is described: Gaussian mixture model (GMM), hidden Markov model (HMM), conditional random fields (CRFs), and dynamic motion primitives (DMPs). A shared underlying concept is that the intent of the demonstrator in performing a task is modeled with a set of latent states, which are inferred from the observed states of the task, that is, the recorded demonstrated trajectories. By utilizing the set of training examples from the step of task perception, the relations between the latent and observed states, as well as the evolution of the states of the model, are derived within a probabilistic framework.

4.1 Gaussian Mixture Model (GMM)

GMM is a parametric probabilistic model for encoding data with a mixture of a finite number of Gaussian probability density functions. For a dataset $\mathbf{Y}$ consisting of multidimensional data vectors $\mathbf{y}_i$, the GMM represented with a mixture of N Gaussian components is (Calinon, 2009)

$$\mathcal{P}(\mathbf{y}_i) = \sum_{n=1}^{N} \mathcal{P}(n)\mathcal{N}(\mathbf{y}_i|n).$$

(4.1)

In (4.1), $\mathcal{P}(n)$, for $n = 1, 2, ..., N$, are the mixture weights for the Gaussian components, calculated as prior probabilities for each component n, i.e.,

$$\mathcal{P}(n) = \pi_n.$$

(4.2)

Robot Learning by Visual Observation, First Edition. Aleksandar Vakanski and Farrokh Janabi-Sharifi.

Similarly, $\mathcal{N}(\mathbf{y}_i|n)$ are conditional probability density functions for the Gaussian components, calculated as

$$\mathcal{N}(\mathbf{y}_i|n) = \frac{1}{\sqrt{(2\pi)^{D+1}|\mathbf{\Sigma}_n|}} e^{-1/2\left[(\mathbf{y}_i - \mathbf{\mu}_n)^T \mathbf{\Sigma}_n^{-1}(\mathbf{y}_i - \mathbf{\mu}_n)\right]}, \qquad (4.3)$$

where $\mathbf{\mu}_n$ and $\mathbf{\Sigma}_n$ are the mean and covariance matrix of the Gaussian component n.

In modeling a set of M demonstrations using GMM, an important requirement is that the sequences of observed data have equal number of data points. Therefore, a preprocessing step of scaling and aligning the demonstration sequences is performed. Methods for trajectories scaling and aligning are presented in Chapter 3, where dynamic temporal warping (DTW) is the most commonly applied method in the preprocessing step in modeling human motions with GMM (Calinon *et al.*, 2007). The input data in modeling human trajectories using GMM is represented by a temporal and spatial component. More specifically, for a D-dimensional measurement vector corresponding to a single trajectory point at time t_k given with $\mathbf{x}_m(t_k) = \left[x_m^{(1)} \; x_m^{(2)} \; \ldots \; x_m^{(D)}\right]$, the input data is a $(D+1)$-dimensional vector that incorporates the time index of the data point as the first element of the vector, that is, $\left[t_k \; x_m^{(1)} \; x_m^{(2)} \; \ldots \; x_m^{(D)}\right]$.

The parameters of a GMM model, π_n, $\mathbf{\mu}_n$, and $\mathbf{\Sigma}_n$, are typically estimated by using the expectation maximization (EM) algorithm (Dempster and Rubin, 1977) in a batch mode. That is, the input dataset consists of concatenated observation data points from all demonstrations. For example, if the demonstration sequences are scaled to a fixed duration of T_m data points in the preprocessing step, then the input set will consist of $T_m M$ vectors with $D+1$ dimensionality. Initial estimates for the model parameters are commonly calculated using k-means clustering. The EM algorithm performs iteratively expectation and maximization steps in order to maximize the estimate of the model parameters, by using a metric, such as the log-likelihood or the posterior.

The expectation step in the EM algorithm defines an expectation function which is evaluated by using the current estimate of the model parameters.

$$\mathcal{P}_{n,i}^{(t_{k+1})} = \frac{\pi_n^{(t_k)} \mathcal{N}\left(\mathbf{y}_i; \mathbf{\mu}_n^{(t_k)}, \mathbf{\Sigma}_n^{(t_k)}\right)}{\sum\limits_{j=1}^{N} \pi_j^{(t_k)} \mathcal{N}\left(\mathbf{y}_i; \mathbf{\mu}_j^{(t_k)}, \mathbf{\Sigma}_j^{(t_k)}\right)}. \qquad (4.4)$$

$$E_n^{(t_{k+1})} = \sum\limits_{j=1}^{N} \mathcal{P}_{n,i}^{(t_{k+1})}. \qquad (4.5)$$

In the maximization step, the model parameters are updated by using the output from the expectation step

$$\pi_n^{(t_k+1)} = \frac{E_n^{(t+1)}}{N}. \tag{4.6}$$

$$\mu_n^{(t_k+1)} = \frac{\sum_{i=1}^{T_mM} \mathcal{P}_{n,i}^{(t_k+1)} \mathbf{y}_i}{E_n^{(t_k+1)}}. \tag{4.7}$$

$$\Sigma_n^{(t_k+1)} = \frac{\sum_{i=1}^{T_mM} \mathcal{P}_{n,i}^{(t_k+1)} \left(\mathbf{y}_i - \mu_n^{(t_k+1)}\right)\left(\mathbf{y}_i - \mu_n^{(t_k+1)}\right)^T}{E_n^{(t_k+1)}}. \tag{4.8}$$

The parameters are iteratively updated in a series of EM steps until satisfaction of a stopping criterion. Most often, a threshold for the log-likelihood of the updated model averaged over the data points is employed (Calinon *et al.*, 2007), i.e.,

$$\frac{\mathcal{L}^{(t_k+1)}}{\mathcal{L}^{(t_k)}} < A_s, \tag{4.9}$$

where

$$\mathcal{L} = \frac{1}{T_mM} \sum_{i=1}^{T_mM} \log(\mathcal{P}(\mathbf{y}_i)). \tag{4.10}$$

4.2 Hidden Markov Model (HMM)

HMM (Rabiner, 1989) represents a doubly stochastic process character-ized by a sequence of hidden states and a sequence of observations. The main idea behind the doubly stochastic processes is that an unobserved stochastic process is affected by an observed stochastic process. In the case of HMM, the unobserved stochastic process has an underlying dependence structure modeled as a Markov chain. A simple schematic representation of an HMM is given in Figure 4.1. The nodes in the graph represent the variables, whereas the edges depict the conditional depen-dencies between the variables. Furthermore, at the time instants t_k for $k = 1, 2, ..., T$, the white nodes represent the hidden states of the HMM denoted with s_k, and the observed variables denoted by o_k are depicted with the shaded nodes.

   HMM has been used extensively in speech recognition (Rabiner and Juang, 1993; Jelinek, 1997), handwriting recognition (Nag *et al.*, 1986),

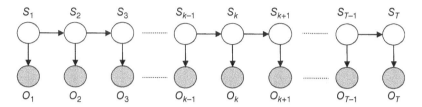

Figure 4.1 Graphical representation of an HMM. The shaded nodes depict the sequence of observed elements $\mathcal{O} = \{o_1, o_2, ..., o_T\}$ and the white nodes depict the hidden states sequence $\mathcal{S} = \{s_1, s_2, ..., s_T\}$.

natural language modeling (Manning and Schutze, 1999), DNA sequence analysis (Durbin *et al.*, 1998), etc.

The Markovian assumption states that given the present hidden state s_k, the future states are independent from the past states

$$\mathcal{P}(s_{k+1}|s_k, s_{k-1}, s_{k-2}, ..., s_1) = \mathcal{P}(s_{k+1}|s_k). \tag{4.11}$$

State transition probabilities are the probabilities of being in state j at time t_{k+1}, given that the model was in state i at time t_k, that is, $a_{i,j} = \mathcal{P}(s_{k+1} = j|s_k = i)$. The state transition probabilities between all hidden states form the state transition matrix $\mathbf{A} = \{a_{i,j}\}$. Since the coefficients $a_{i,j}$ represent probabilities, they satisfy the standard probabilistic requirements $a_{i,j} \geq 0$ and $\sum_j a_{i,j} = 1$. HMMs assume that state transition probabilities are constant over time for all observations.

The probabilities of the model being at state i at time t_1 form a matrix of initial state probabilities $\boldsymbol{\pi} = \{\pi_i = \mathcal{P}(s_1 = i)\}$ Subsequently, $\pi_i \geq 0$ and $\sum \pi_i = 1$.

For a set of N_o observation symbols $\{c_1, c_2, ..., c_{N_o}\}$, the probabilities of observing the symbol c_f at time t_k and the model is in state i, that is, $b_i(f) = \mathcal{P}(o_k = c_f|s_k = i)$, form the observation probability matrix $\mathbf{B} = \{b_i(f)\}$. The coefficients obey the properties $b_i(f) \geq 0$ and $\sum b_i(f) = 1$.

Another important assumption of HMMs is the temporal independence of the observations, which is also reflected in the graphical model in Figure 4.1 by the lack of edges between the observed (shaded) nodes.

HMMs are fully described by the sets of matrices $\boldsymbol{\pi}$, $\mathbf{A}$, and $\mathbf{B}$. Usually, a compact notation for an HMM is used that combines the three matrices $\lambda = \{\boldsymbol{\pi}, \mathbf{A}, \mathbf{B}\}$.

The main problems solved with HMM are referred to as evaluation, decoding, and training.

4.2.1 Evaluation Problem

Evaluation concerns the problem of finding $\mathcal{P}(\mathcal{O}|\lambda)$, which is the probability of outputting a sequence of observations $\mathcal{O} = (o_1, o_2, ..., o_{k-1}, o_k, o_{k+1}, ..., o_T)$ for a given model $\lambda = \{\pi, \mathbf{A}, \mathbf{B}\}$. Two approaches are employed for solving the evaluation problem, known as the forward algorithm and the backward algorithm (Baum and Egon, 1967).

The forward algorithm consists of introducing forward variables $\bar{\alpha}_k(i) = \mathcal{P}(o_1, o_2, ..., o_{k-1}, o_k, s_k = i|\lambda)$, defined as the probabilities of observing the partial sequence $o_1, o_2, ..., o_{k-1}, o_k$ until time t_k and being at state i, given the model λ. The steps of recursively calculating the probabilities are as follows:

1) Initialization: $\overline{\alpha_1}(i) = \pi_i b_i(o_1)$, for $1 \le i \le N_s$.

2) Induction: $\bar{\alpha}_{k+1}(j) = \left[\sum_{i=1}^{N_s} \bar{\alpha}_k(i) a_{i,j} \right] b_j(o_{k+1})$, for $1 \le j \le N_s$ and $k = 1, 2, ..., T-1$.

3) Termination: $\mathcal{P}(\mathcal{O}|\lambda) = \sum_{i=1}^{N_s} \bar{\alpha}_T(i)$.

The notation N_s is used to denote the total number of hidden states of an HMM.

The backward algorithm is analogous to the forward algorithm and it is based on introducing backward variables $\bar{\beta}_k(i) = \mathcal{P}(o_{k+1}, o_{k+2}, ..., o_{T-1}, o_T|s_k = i, \lambda)$, which represent the probabilities of observing the partial sequence $o_{k+1}, o_{k+2} ..., o_{T-1}, o_T$, from time t_{k+1} until the end time T, given that the state at time t_k is i and the model λ is known. The recursive procedure for the backward algorithm comprises the following:

1) Initialization: $\bar{\beta}_T(i) = 1$, for $1 \le i \le N_s$.

2) Induction: $\bar{\beta}_k(i) = \sum_{j=1}^{N_s} a_{i,j} b_j(o_{k+1}) \bar{\beta}_{k+1}(j)$, for $1 \le j \le N_s$ and $k = T-1, T-2, ..., 1$.

3) Termination: $\mathcal{P}(\mathcal{O}|\lambda) = \sum_{i=1}^{N_s} \pi_i b_i(o_1) \bar{\beta}_1(i)$.

Either forward or backward approach can be used for solving the evaluation problem. For given different HMMs $\lambda_1 = \{\pi_1, \mathbf{A}_1, \mathbf{B}_1\}$ and $\lambda_2 = \{\pi_2, \mathbf{A}_2, \mathbf{B}_2\}$, these algorithms can be used to find the model that better explains an observed sequence. Based on learned models of human actions from demonstrations, HMMs can recognize human actions through classification into one of the learned models. In a PbD setting, the skill acquisition process can be accomplished as learning and recognition of basic human actions.

4.2.2 Decoding Problem

Decoding relates to the problem of finding a sequence of states $s_1, s_2, ..., s_T$ that maximizes the probability of observing a sequence $o_1, o_2, ..., o_T$ for a given model λ, or $\mathcal{P}(\mathcal{S}|\mathcal{O},\lambda)$. A solution is proposed by Viterbi (1967).

The Viterbi algorithm introduces a new variable

$$\bar{\delta}_k(i) = \max_{s_1, s_2, ..., s_{k-1}} \mathcal{P}(s_1, s_2, ..., s_{k-1}, s_k = i, o_1, o_2, ..., o_k|\lambda), \qquad (4.12)$$

which refers to the maximum likelihood of a sequence of states up to time t_k and state i at time t_k, and the first k observations given the model. The state at time t_{k-1} that maximizes the likelihood of transition to state i at time t_k is denoted as $\bar{\psi}_1(i) = 0$. The algorithm consists of the following steps:

1) Initialization: $\bar{\delta}_1(i) = \pi_i b_i(o_1)$, $\bar{\psi}_1(i) = 0$, for $1 \le i \le N_s$.
2) Recursion: $\bar{\delta}_k(j) = \max_{1 \le i \le N_s} \left[\bar{\delta}_{k-1}(i)a_{i,j}\right]b_j(o_k)$,

$\bar{\psi}_k(j) = \arg \max_{1 \le i \le N_s} \left[\bar{\delta}_{k-1}(i)a_{i,j}\right]$, for $2 \le k \le T$, $1 \le j \le N_s$.

3) Termination: $s_T^* = \arg \max_{1 \le i \le N_s} \left[\bar{\delta}_T(i)\right]$.
4) State sequence backtracking: $s_T^* = \bar{\psi}_{k+1}\left(s_{k+1}^*\right)$, for $k = T-1, T-2, ..., 1$.

Note that an exact solution to the decoding problem does not exist; instead, the obtained solution for the states sequence is optimal with respect to a particular optimality criterion.

4.2.3 Training Problem

Training a model involves estimating the model parameters $\lambda = \{\pi, \mathbf{A}, \mathbf{B}\}$ that maximize the probability of the observed data. Analytical solution to the training problem is not available. The Baum–Welch algorithm (Baum *et al.*, 1970) is often employed for identifying the model parameters, by locally maximizing the likelihood of the training data $\mathcal{P}(\mathcal{O}|\lambda)$ in an iterative procedure.

The algorithm introduces a variable which denotes the probability of being in state i at time t_k and making a transition to state j at time t_{k+1}, given the observation sequence and the model, that is, $\bar{\xi}_k(i,j) = \mathcal{P}(s_k = i, s_{k+1} = j|o_1, o_2, ..., o_T, \lambda)$. Using Bayes theorem, it can be rewritten as a function of the forward and backward variables $\bar{\alpha}$ and $\bar{\beta}$, i.e.,

$$\bar{\xi}_k(i,j) = \frac{\bar{\alpha}_k(i)a_{i,j}b_j(o_{k+1})\bar{\beta}_{k+1}(j)}{\mathcal{P}(o_1,o_2,...,o_T)}. \tag{4.13}$$

In addition, another variable is defined as the probability of being at state i at time t_k, given the observation sequence and the model,

$$\bar{\theta}_k(i) = \mathcal{P}(s_k = i|\mathcal{O},\lambda) = \frac{\bar{\alpha}_k(i)\bar{\beta}_k(i)}{\mathcal{P}(\mathcal{O}|\lambda)}. \tag{4.14}$$

The Baum–Welch parameter estimation formulas are as follows:

$$\hat{\pi}_i = \text{expected number of times at state } i \text{ at time } 1 = \bar{\theta}_1(i),$$

$$\hat{a}_{i,j} = \frac{\text{expected number of transitions from state } i \text{ to state } j}{\text{expected number of transitions from state } i}$$

$$= \frac{\sum_{k=1}^{T-1} \bar{\xi}_k(i,j)}{\sum_{k=1}^{T-1} \bar{\theta}_k(i)},$$

$$\hat{b}_j(f) = \frac{\text{expected number of times in state } j \text{ and observing symbol } c_f}{\text{expected number of times in state } j}$$

$$= \frac{\sum_{\substack{k=1 \\ o_k = c_f}}^{T} \bar{\theta}_k(j)}{\sum_{k=1}^{T} \bar{\theta}_k(j)}. \tag{4.15}$$

For the estimated model $\hat{\lambda} = \{\hat{\pi},\hat{\mathbf{A}},\hat{\mathbf{B}}\}$ from (4.15), and a given initial model λ, it was proven in the work of Baum and Seli (1968) that either the initial model is a critical point of the likelihood function (in which case $\hat{\lambda} = \lambda$), or the likelihood of the observation sequence given the re-estimated model is greater when compared to the initial model, that is, $\mathcal{P}(\mathcal{O}|\hat{\lambda}) \geq \mathcal{P}(\mathcal{O}|\lambda)$. This procedure is iterated until a stopping criterion is satisfied. The solution is referred to as a maximum likelihood estimate of the HMM, since its goal is to find a set of parameters which maximizes the likelihood of the observation data.

4.2.4 Continuous Observation Data

In the earlier discussion, it was considered that the observations are discrete symbols chosen from a finite set of elements. In cases when the observations are continuous, one alternative is to quantize the signals into a discrete codebook of symbols. The other alternative is to use continuous observation densities. A general form of a probability density function is a finite mixture

$$b_i(\mathcal{O}) = \sum_{m=1}^{N_m} \bar{g}_{i,m}\, \mathcal{M}\{\mathcal{O}, \bar{\boldsymbol{\mu}}_{i,m}, \bar{\boldsymbol{\Sigma}}_{i,m}\}, \tag{4.16}$$

where $\bar{g}_{i,m}$ is the mixture coefficient for the mth mixture in state i, and $\mathcal{M}$ is any log-concave or elliptically symmetric density function with mean vector $\bar{\boldsymbol{\mu}}_{i,m}$ and covariance matrix $\bar{\boldsymbol{\Sigma}}_{i,m}$ for the mth mixture in state i. The mixture coefficients $\bar{g}_{i,m}$ satisfy the stochastic constraints $\sum_{m=1}^{N_m} \bar{g}_{i,m} = 1$ and $\bar{g}_{i,m} \geq 0$ for $1 \leq i \leq N_s$ and $1 \leq m \leq N_m$. Continuous observation signals in HMMs are most often represented with a mixture of Gaussian density functions. For further implementation details imposed by the use of continuous observation densities in HMMs, see Rabiner (1989).

4.3 Conditional Random Fields (CRFs)

CRF (Lafferty *et al.*, 2001) is a discriminative probabilistic approach for finding the conditional probability distribution of a sequence of hidden states for a given sequence of observations. CRF belongs to the family of undirected graphical models (Bishop, 2006). The graphical structure of CRFs can be very complex in general. However, since sequential data are considered in the book, related to observed demonstrated trajectories at defined time instances, the configuration of graphical models in a form of a linear chain is considered. The linear chain CRF structure is displayed in Figure 4.2. Another example of a graphical model with linear chain architecture is HMM (see Figure 4.1). Different from CRFs, HMMs belong to the family of directed graphical models, in which the output

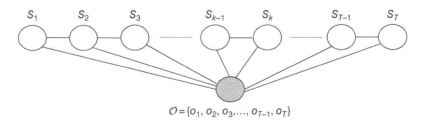

$\mathcal{O} = \{o_1, o_2, o_3, \ldots, o_{T-1}, o_T\}$

Figure 4.2 Graphical representation of a CRF with linear chain structure.

nodes precede the input nodes in the graph. Or, in other words, in a directed graphical model, the outputs are probabilistically generated by the inputs, and therefore these models are also called generative models.

A subset of nodes in an undirected graph that contains a link between each pair of nodes is called *clique*. The cliques with the maximum number of conditionally dependent variables are the *maximal cliques*. In CRFs, the joint probability distribution of the graph variables is defined as a product of potential functions ψ_C, with each potential function corresponding to a maximal clique in the graph over a subset of variables $\mathbf{y}_C$, that is,

$$P(\mathbf{y}) = \frac{1}{W}\prod_C \psi_C(\mathbf{y}_C). \tag{4.17}$$

The quantity W is called normalization factor (or partition function), and ensures that $P(\mathbf{y})$ in (4.17) represent a properly normalized distribution, i.e.,

$$W = \sum_{\mathbf{y}}\prod_C \psi_C(\mathbf{y}_C). \tag{4.18}$$

In addition, in order to have proper values for the probabilities in (4.17), the potential functions $\psi_C(\mathbf{y}_C)$ must be real valued and strictly positive. Consequently, the potential functions are usually expressed as exponentials of energy functions, that is, $\psi_C(\mathbf{y}_C) = \exp\{\varepsilon(\mathbf{y}_C)\}$. As a result, the joint distribution (4.17) is represented as a sum of the energies of maximal cliques in the graph

$$P(\mathbf{y}) = \frac{1}{W}\exp\left(\sum_C \varepsilon(\mathbf{y}_C)\right). \tag{4.19}$$

4.3.1 Linear Chain CRF

In the case of a linear chain CRF, the maximal cliques comprise a pair of adjacent hidden state variables and the sequence of observations $\mathcal{O}$. For example, for the hidden state variable s_k, the maximal clique is shown in the middle in Figure 4.2. It includes the link between the hidden states (s_k, s_{k-1}), and the links connecting the hidden states s_k and s_{k-1} with the observation sequence. The assumption of first-order Markov property, given in (4.11), holds for the sequence of state variables in the CRF graph structure, that is, each state is conditionally independent of the other states given its neighbor's states.

For a given observations sequence $\mathcal{O}$ and a sequence of unobserved states $\mathcal{S}$, a linear chain CRF is defined as a conditional distribution of the states given the observations (Lafferty *et al.*, 2001; Sutton and McCallum, 2006)

$$P(\mathcal{S}|\mathcal{O}) = \frac{1}{W}\exp\left\{\sum_{l_1}\bar{\varphi}_{l_1}\Phi_{l_1}(s_k,s_{k-1},\mathcal{O},t_k) + \sum_{l_2}\bar{\gamma}_{l_2}\Gamma_{l_2}(s_k,\mathcal{O},t_k)\right\},$$

(4.20)

where the notation $\Phi_{l_1}(s_k,s_{k-1},\mathcal{O},t_k)$ is used for the *transition feature functions* of the hidden states at times t_k and t_{k-1} and the observation sequence $\mathcal{O}$, whereas $\Gamma_{l_2}(s_k,\mathcal{O},t_k)$ denote *state feature functions* relating the observation sequence $\mathcal{O}$ and the hidden state at time t_k. The symbols l_1 and l_2 are employed for indexing the transition and state feature functions, respectively. The set $\Lambda = \{\bar{\varphi}_{l_1},\bar{\gamma}_{l_2}\}$ for $l_1 = 1,2,\dots$ and $l_2 = 1,2,\dots,$ denotes the parameters of the model. Often the representation of CRFs is simplified by generalizing the notation of the aforementioned two feature functions into $\Theta(s_k,s_{k-1},\mathcal{O},t_k)$. Using this notation, $\Theta(s_k,s_{k-1},\mathcal{O},t_k)$ can represent either a transition feature function $\Phi(s_k,s_{k-1},\mathcal{O},t_k)$ or a state feature function $\Gamma(s_k,\mathcal{O},t_k)$, and the model (4.20) can thus be written in the form

$$P(\mathcal{S}|\mathcal{O}) = \frac{1}{W}\exp\left\{\sum_{j}\theta_j\Theta_j(s_k,s_{k-1},\mathcal{O},t_k)\right\},$$

(4.21)

where the notation θ_j is used for the model parameters associated to the unified feature functions Θ_j, for $j = 1,2,\dots$. As noted earlier, the normalization function W is obtained by summing over all possible configurations on $\mathcal{S}$, and enforces that the cumulative probabilities sum to one

$$W = \sum_{s}\exp\left\{\sum_{j}\theta_j\Theta_j(s_k,s_{k-1},\mathcal{O},t_k)\right\}.$$

(4.22)

The choice of feature functions Θ_j is application dependent. Several possible variants for the feature functions in trajectory learning will be discussed in Section 5.2.3.

4.3.2 Training and Inference

The problem of estimation of the parameters of the model $\Lambda = \{\theta_j\}$ is typically solved by maximizing the conditional likelihood of the training data. In fact, the maximization is performed by using the natural logarithm of the likelihood function (referred to as log-likelihood), since it facilitates the procedure. For a given independent and identically distributed training set $\{\mathcal{O}_m,\mathcal{S}_m\}_{m=1}^{M}$ consisting of M observation sequences $\mathcal{O}$ and the corresponding state sequences $\mathcal{S}_m$, the log-likelihood is obtained as follows:

$$\mathcal{L}(\Lambda) = \sum_{m=1}^{M} \log \mathcal{P}(\mathcal{S}_m | \mathcal{O}_m) = \sum_{m=1}^{M} \left[\sum_{j=1}^{J} \theta_j \Theta_j(s_{k,m}, s_{k-1,m}, \mathcal{O}, t_k) - \log W(\mathcal{O}_m) \right].$$

$$(4.23)$$

A noticeable property of the function $\mathcal{L}(\Lambda)$ in (4.23) is that it represents a logarithm of sum of exponential components, and as a consequence it is concave. This property significantly facilitates the problem of likelihood maximization and guarantees that any found local maximum is a global maximum.

For the partial derivatives of the log-likelihood in (4.23) with respect to model parameters $\Lambda = \{\theta_j\}$, one obtains

$$\frac{\partial \mathcal{L}}{\partial \theta_j} = \sum_{m=1}^{M} \Theta_j(s_{k,m}, s_{k-1,m}, \mathcal{O}, t_k)$$

$$- \sum_{m=1}^{M} \sum_{\mathcal{S}} \Theta_j(s_{k,m}, s_{k-1,m}, \mathcal{O}, t_k) \mathcal{P}(s_{k,m}, s_{k-1,m} | \mathcal{O}_m, t_k).$$

$$(4.24)$$

Setting the partial derivatives in (4.24) to zero results in the maximum entropy solution, that is, the expectations of each feature function Θ_j with respect to the empirical distribution of the training data and the expectations of the feature functions with respect to the model distribution are equalized.

Once the log-likelihood function (4.23) and its derivatives (4.24) are obtained, estimation of the model parameters is solved by numerical optimization methods. Often quasi-Newton methods with approximation of the Hessian matrix are employed, such as limited-memory Broyden–Fletcher–Goldfarb–Shanno (BFGS) algorithm (Sutton and McCallum, 2006) or conjugate gradient approach (Lafferty *et al.*, 2001).

Over-fitting of the model parameters can often be avoided by regularization of the optimization function (4.23). In that case, the log-likelihood is penalized by adding an additional term proportional to the Euclidean norm of the parameters vector $\sum \theta_j^2$, or by adding an additional term proportional to the l_1-norm of the parameters $\sum |\theta_j|$ (Vail *et al.*, 2007).

Inference in linear chain CRFs is associated with solving two problems. The first one is finding the most likely sequence of states for a given sequence of observations and known model parameters. The second problem is related to finding the pairwise marginal probabilities of the states $\mathcal{P}(s_k, s_{k-1} | \mathcal{O}, t_k)$, which are required for computation of the partial derivatives in (4.24) for parameter estimation during the training. These two problems are efficiently solved with the Viterbi and the forward–backward algorithm, respectively (Lafferty *et al.*, 2001).

4.4 Dynamic Motion Primitives (DMPs)

DMPs in learning from demonstrations refers to a body of works which is oriented toward learning attractor landscapes based on the dynamics of demonstrated trajectories, as opposed to the PbD methods directed toward generating a single generalized trajectory (Billard *et al.*, 2008). The dynamical systems exploit the observed information for the positions, velocities, and/or accelerations of the demonstrations, and approximate the nonlinear dynamic of the motion with mixtures of linear systems. The pioneering work in Ijspeert *et al.* (2001) employed a set of differential equations for modeling demonstrated trajectories, which was coupled with a canonical dynamical system for controlling the temporal dynamics of the system (Ijspeert *et al.*, 2002). The initial DMPs model for encoding task with discrete goals was later extended for periodic tasks characterized with a defined phase of oscillation. Several later works (Gams and Ude, 2009; Hoffmann *et al.*, 2009; Pastor *et al.*, 2009) proposed certain modifications or improvements of the model. Another group of researchers working on learning the motion dynamics put the main emphasis on modeling the multivariate joint probability distribution of the demonstrated positions, velocities, and/or accelerations. Statistical parametric models, such as HMM (Calinon *et al.*, 2010) or GMM (Gribovskaya *et al.*, 2010; Khansari-Zadeh and Billard, 2010), were used to encode the correlations between the dynamical variables of the demonstrated trajectories and to generate robot reproduction policies.

The dynamical systems approach formulated in the work of Ijspeert *et al.* (2002) and its modifications (Gams and Ude, 2009; Hoffmann *et al.*, 2009; Pastor *et al.*, 2009) use a mass–damper–spring model

$$M_d \ddot{x} + B_d \dot{x} + K_d \left(x_{\text{goal}} - x \right) = F, \tag{4.25}$$

where x_{goal} denotes the desired steady-state value of the state variable, M_d, B_d, and K_d are the parameters of the model, and F is a nonlinear forcing term which is learned from the demonstrated motions.

The dynamical system (4.25) is often written in a state-space form $\dot{\mathbf{x}} = f(\mathbf{x}, \varphi, \theta)$, with states $\mathbf{x} = [x_1 \ x_2]$, a canonical variable φ, and a set of high-level parameters θ learned from the demonstrations, such as period, baseline, and amplitude, in the case of periodic motions. Thus, (4.25) can be rewritten as

$$\tau \dot{x}_2 = \alpha \left(\beta \left(x_{\text{goal}} - x_1 \right) - x_2 \right) + F,$$
$$\tau \dot{x}_1 = x_2,$$
(4.26)

where x_1 and x_2 denote the position and velocity of the trajectories, respectively, and τ is a constant scaling factor of the movement duration. For the case of periodic motions, τ defines the period of the motions. The parameters α and β are chosen so that the system is critically damped (in order to avoid overshooting the desired values of the variables). The force term F is a nonlinear periodic function of the canonical variable φ. The variable φ defines a canonical system which monotonically changes between 1 and 0 for the duration of the movement, and it is calculated as

$$\tau \dot{\varphi} = -\alpha \varphi.$$
(4.27)

In the case of periodic movement, the canonical variable φ anchors the dynamical system into the phase of the oscillations, and is often formulated as a phase oscillator

$$\dot{\varphi} = \omega,$$
(4.28)

where ω is the fundamental frequency component of the oscillations. Other types of oscillators can be used for the canonical system (4.28) as well.

The forcing term F is represented as a linear combination of a set of exponential basis functions

$$F = \frac{\sum_{i=1}^{N} w_i \Psi_i}{\sum_{i=1}^{N} \Psi_i} \varphi,$$
(4.29)

where Ψ_i denote the N basis function. The weights w_i are adjusted to fit arbitrary nonlinear functions via locally weighted linear regression in either a batch mode or an incremental mode.

Typically, Gaussian basis functions are used for learning discrete tasks in the form

$$\Psi_i = e^{-h_i \, (\varphi_i - c_i)^2},$$
(4.30)

defined by their width h_i and centers c_i.

For learning periodic motions, the following von Mises basis functions are usually adopted:

$$\Psi_i = e^{-(1/2)h_i(1 - \cos(\varphi_i - c_i))},$$
(4.31)

where, similarly, h_i and c_i denote widths and centers, respectively. They are usually uniformly distributed within one period of the motion (e.g., in the range of $-\pi$ to π).

The dynamical system (4.26) has a stable global attractor point at the goal of the motions $[x_1 \ x_2] \rightarrow [x_{\text{goal}} \ 0]$, and the nonlinear periodic force F vanishes at the end of the discrete movement, thus ensuring convergence of the motion. In the case of periodic movements, the global attractor point is the baseline of the movements, whereas the force produces limit cycle oscillators.

The main advantage of using dynamical systems for robot learning consists of employing stable models which drive the system toward the attractor from different initial conditions. The learned motions are robust to perturbations and parameters variations, which is important for reproduction of learned skills in unstructured environments. Moreover, the learned motions are spatially and temporally invariant, meaning that similar trajectories with different goal, starting point, or temporal scaling, can be reproduced from a learned model of the discrete motions, or for periodic motions accordingly, trajectories with different amplitude, baseline, and frequency can be reproduced.

4.5 Summary

The chapter describes the most common methods used for task modeling at the low-level of abstraction. GMM employs a mixture of Gaussian probability distributions for modeling a task represented with multiple demonstrated trajectories. HMM is a generative probabilistic approach that uses directed connections between a set of hidden and observed states for task modeling. The problems of evaluation, decoding and training in HMM are reviewed. CRF, on the other hand, is a discriminative probabilistic approach belonging to the family of undirected graphical models. An overview of training and inference in linear chain CRF, related to trajectory modeling, is presented. The last discussed method, DMPs, is based on the dynamical systems formalism and employs a set of differential equations in modeling the dynamics of perceived motions. A force component is learned from the training trajectories, and drives the temporal evolution of the system ensuring convergence toward either a discrete task goal, or toward a limit cycle oscillator in the case of tasks with periodic motions.

References

Baum, L., and Egon, J.A., (1967). An inequality with applications to statistical estimation for probabilistic functions of Markov process and to a model for ecology. *Bulletin of the American Meteorological Society*, vol. **73**, no. 3, pp. 360–362.

Baum, L.E., and Seli, G.R., (1968). Growth functions for transformations on manifolds. *Pacific Journal of Mathematics*, vol. **27**, no. 2, pp. 211–227.

Baum, L., Petrie, T., Soules, G., and Weiss, N., (1970). A maximization technique occurring in the statistical analysis of probabilistic functions of Markov chains. *The Annals of Mathematical Statistics*, vol. **41**, no. 1, pp. 164–171.

Billard, A., Calinon, S., Dillmann, R., and Schaal, S., (2008). Robot programming by demonstration. *Handbook of Robotics*. (Eds.) Siciliano, B., and Oussama, K., Berlin, Germany: Springer-Verlag, ch. 59.

Bishop, C.M., (2006). *Pattern Recognition and Machine Learning*. New York, USA: Springer.

Calinon, S., (2009). *Robot Programming by Demonstration: A Probabilistic Approach*. Boca Raton, USA: EPFL/CRC Press.

Calinon, S., Guenter, F., and Billard, A., (2007). On learning, representing and generalizing a task in a humanoid robot. *IEEE Transaction on Systems, Man and Cybernetic—Part B*, vol. **37**, no. 2, pp. 286–298.

Calinon, S., D'halluin, F., Sauser, E.L., Caldwell, D.G., and Billard, A.G., (2010). Learning and reproduction of gestures by imitation: an approach based on hidden Markov model and Gaussian mixture regression. *IEEE Robotics and Automation Magazine*, vol. **17**, no. 2, pp. 44–54.

Dempster, A, and Rubin, N.L.D., (1977). Maximum likelihood from incomplete data via the EM algorithm. *Journal of the Royal Statistical Society*, vol. **39**, no. 1, pp. 1–38.

Durbin, R., Eddy, S., Krogh, A., and Mitchison, G., (1998). *Biological Sequence Analysis: Probabilistic Models of Proteins and Nucleic Acids*. Cambridge, USA: Cambridge University Press.

Gams, A., and Ude, A., (2009). Generalization of example movements with dynamic systems. *Proceedings of IEEE/RAS International Conference on Humanoid Robots*, Paris, France, pp. 28–33.

Gribovskaya, E., Khansari-Zadeh, S.M., and Billard, A., (2010). Learning nonlinear multivariate dynamics of motion in robotics manipulators. *International Journal of Robotics Research*, vol. **30**, no. 1, pp. 80–117.

Hoffmann, H., Pastor, P., Park, D.-H., and Schaal, S., (2009). Biologically-inspired dynamical systems for movement generation: automatic real-time goal adaptation and obstacle avoidance. *Proceedings on International Conference on Robotics and Automation*, Kobe, Japan, pp. 2587–2592.

Ijspeert, A.J., Nakanishi, J., and Schaal, S., (2001). Trajectory formation for imitation with nonlinear dynamical systems. *Proceedings of IEEE/RSJ International Conference on Intelligent Robots Systems*, Maui, USA, 2001, pp. 752–757.

Ijspeert, A.J., Nakanishi, J., and Schaal, S., (2002). Movement imitation with nonlinear dynamical systems in humanoid robots. *Proceedings of 2002 IEEE*

International Conference on Robotics and Automation, Washington, USA, pp. 1398–1403.

Jelinek, F., (1997). *Statistical Methods for Speech Recognition*. Cambridge, USA: MIT Press.

Khansari-Zadeh, S.M., and Billard, A., (2010). BM: An iterative algorithm to learn stable non-linear dynamical systems with Gaussian mixture models. *Proceedings of International Conference on Robotics and Automation*, Anchorage, USA, pp. 2381–2388.

Lafferty, J., McCallum, A., and Pereira, F., (2001). Conditional random fields: probabilistic models for segmenting and labeling sequence data. *Proceedings of International Conference on Machine Learning*, Williamstown, USA, pp. 282–289.

Manning, C.D., and Schutze, H., (1999). *Foundations of Statistical Natural Language Processing*. Cambridge, USA: MIT Press.

Nag, R., Wong, K., and Fallside, F., (1986). Script recognition using hidden Markov models. *Proceedings of IEEE International Conference on Acoustics, Speech and Signal Processing*, Tokyo, Japan, pp. 2071–2074.

Pastor, P., Hoffmann, H., Asfour, T., and Schaal, S., (2009). Learning and generalization of motor skills by learning from demonstration. *Proceedings of International Conference on Robotics and Automation*, Kobe, Japan, pp. 763–768.

Rabiner, L., (1989). A tutorial on hidden Markov models and selected applications in speech recognition. *Proceedings of the IEEE*, vol. **77**, no. 2, pp. 257–286.

Rabiner, L., and Juang, B.H., (1993). *Fundamentals of Speech Recognition*. Upper Saddle River, USA: Prentice Hall.

Sutton, C., and McCallum, A., (2006). *An Introduction to Conditional Random Fields for Relational Learning*. (Eds.) Getoor, L., and Taskar, B., Cambridge, USA: MIT Press.

Vail, D.L., Lafferty, J.D., and Veloso, M.M., (2007). Feature selection in conditional random fields for activity recognition. *Proceedings of IEEE/RSJ International Conference on Intelligent Robots and Systems*, San Diego, USA, pp. 3379–3384.

Viterbi, A.J., (1967). Error bounds for convolutional codes and an asymptotically optimal decoding algorithm. *IEEE Transactions on Information Theory*, vol. **IT-13**, no. 2, pp. 260–269.

5

Task Planning

Task planning in the context of trajectory learning is related to generation of a plan for reproduction of a demonstrated task. The task planning approaches presented in this chapter employ the task modeling methodology reviewed in Chapter 4. Gaussian mixture regression (GMR) is used for tasks modeled with a Gaussian mixture model (GMM) (Section 4.1). Spline regression is employed with tasks represented with a set of salient trajectory key points. Hidden Markov models (HMMs) (Section 4.2) and conditional random fields (CRFs) (Section 4.3) are commonly used for extracting trajectory key points where the task planning step is performed with spline regression. Locally weighted regression is applied for generation of a reproduction strategy with tasks modeled with the dynamic motion primitives (DMPs) approach (Section 4.4).

5.1 Gaussian Mixture Regression

GMR is used for generating a generalized trajectory for reproduction of tasks represented mathematically with a GMM. As explained in Section 4.1, GMM encodes parametrically a set of observed trajectories with multiple Gaussian probability density functions. For n number of Gaussian components, the model parameters are π_n, μ_n and Σ_n. GMR in this case is employed to find the conditional expectation of the temporal components of observed trajectories given the spatial component of observed trajectories. Or, for a temporal vector indexing the observation times in the recorded sequences denoted by t_k for $k = 1, 2, ..., T_m$, and corresponding D-dimensional measurements $\mathbf{x}_m(t_k) = \begin{bmatrix} x_m^{(1)} & x_m^{(2)} & ... & x_m^{(D)} \end{bmatrix}$ for $k = 1, 2, ..., T_m$, the regression calculates $\mathbb{E}\{\mathcal{P}(\mathbf{x}_m(t_k)|t_k)\}$, that is, the conditional expectation of $\mathbf{x}_m(t_k)$ given t_k.

Robot Learning by Visual Observation, First Edition. Aleksandar Vakanski and Farrokh Janabi-Sharifi.
© 2017 John Wiley & Sons, Inc. Published 2017 by John Wiley & Sons, Inc.

In encoding the trajectories using GMM $(D + 1)$-dimensional input data were used, with the temporal and spatial components of the model parameters for the Gaussian component n denoted by

$$\boldsymbol{\mu}_n = \begin{bmatrix} \boldsymbol{\mu}_t^n & \boldsymbol{\mu}_x^n \end{bmatrix}, \boldsymbol{\Sigma}_n = \begin{bmatrix} \boldsymbol{\Sigma}_{tt}^n & \boldsymbol{\Sigma}_{tx}^n \\ \boldsymbol{\Sigma}_{xt}^n & \boldsymbol{\Sigma}_{xx}^n \end{bmatrix}. \tag{5.1}$$

It is known that the conditional probability of Gaussian variables is also a Gaussian density function. This property is exploited to represent the expected conditional probability $\mathcal{P}(\mathbf{x}_m|\mathbf{t},n)$ as

$$\mathcal{P}(\mathbf{x}_m|\mathbf{t},n) = N\left(\hat{\boldsymbol{\mu}}_x^n, \hat{\boldsymbol{\Sigma}}_{xx}^n\right) \tag{5.2}$$

where

$$\hat{\boldsymbol{\mu}}_x^n = \boldsymbol{\mu}_x^n + \boldsymbol{\Sigma}_{xt}^n \left(\boldsymbol{\Sigma}_{tt}^n\right)^{-1}\left(\mathbf{t} - \boldsymbol{\mu}_t^n\right), \tag{5.3}$$

$$\hat{\boldsymbol{\Sigma}}_{xx}^n = \boldsymbol{\Sigma}_{xx}^n - \boldsymbol{\Sigma}_{xt}^n \left(\boldsymbol{\Sigma}_{tt}^n\right)^{-1}\boldsymbol{\Sigma}_{tx}^n. \tag{5.4}$$

If the weight for the contribution of the Gaussian mixture n is calculated using

$$\lambda_n = \mathcal{P}(n|\mathbf{t}) = \frac{\pi_n \mathcal{P}(\mathbf{t}|n)}{\sum_{j=1}^{N} \mathcal{P}(j)p(\mathbf{t}|j)} = \frac{\pi_n \mathcal{N}\left(\boldsymbol{\mu}_t^n, \boldsymbol{\Sigma}_t^n\right)}{\sum_{j=1}^{N} \mathcal{P}(j)\mathcal{N}\left(\boldsymbol{\mu}_t^j, \boldsymbol{\Sigma}_t^j\right)} \tag{5.5}$$

then estimation of conditional probability distribution for the mixture model is obtained from

$$\hat{\boldsymbol{\mu}}_x = \sum_{j=1}^{N} \lambda_j \hat{\boldsymbol{\mu}}_x^j, \hat{\boldsymbol{\Sigma}}_{xx} = \sum_{j=1}^{N} \lambda_j \hat{\boldsymbol{\Sigma}}_{xx}^n. \tag{5.6}$$

Calculating $\hat{\boldsymbol{\mu}}_x$ at each time step t_k results in a generalized trajectory for task reproduction. The covariance matrices $\hat{\boldsymbol{\Sigma}}_{xx}$ describe the spatial constraints for the resultant trajectory.

The trade-off between the accuracy and the smoothness of the generalized trajectory is controlled by the number of Gaussian components used. Increasing the number of Gaussians improves the accuracy of the model fitting the input data up to a certain level over which overfitting occurs, resulting in nonsmooth generalized trajectory.

5.2 Spline Regression

Spline regression is applied in trajectory learning approaches which rely on task representation using key points. This section describes approaches for extraction of key points from acquired trajectories by employing HMMs and CRFs, followed by task generalization using spline regression.

5.2.1 Extraction of Key Points as Trajectories Features

The statistical methods that use discrete states for modeling human demonstrations (e.g., HMMs) typically involve segmentation of the continuous perceptual data into a sequence of discrete subgoals. The subgoals are key points along the trajectories, often associated with notable changes in the kinematic or dynamic parameters of the trajectories. Hence, the key points should represent events in the demonstrated motions that are important for task reproduction. Accordingly, the task analysis step of such programming by demonstration (PbD) techniques requires identification of a set of candidate key points. Sets of criteria for selection of candidate key points for continuous trajectories have been proposed in the literature (Calinon and Billard, 2004; Asfour *et al.*, 2006). The research presented in the book introduces a different technique for selecting candidate key points, based on classification of discretely labeled data consisting of normalized position and velocity feature vectors (Vakanski *et al.*, 2012). The approach utilizes the Linde–Buzo–Gray (LBG) algorithm (Linde *et al.*, 1980) for segmentation of demonstrated trajectories associated with significant changes in position and velocity. The implementation of the LBG approach for trajectories segmentation was inspired by the works on human motion identification and classification that employ clustering techniques (Liu *et al.*, 2005; Zhou *et al.*, 2008).

The LBG algorithm is a variant of the k-means clustering method, that is, it is used to partition a set of input data points into a number of clusters. A distinct property of the LBG algorithm is the partitioning procedure, which starts with one data cluster followed by iteratively doubling the number of clusters until termination conditions are satisfied. A pseudo-code of the algorithm is given in Table 5.1.

For the problem at hand, the velocity values for each dimension d of the recorded trajectories are calculated using an empirically chosen delay value of ς sampling periods, i.e.,

$$v_m^{(d)}(t_k) = \frac{x_m^{(d)}(t_{k+\varsigma}) - x_m^{(d)}(t_{k-\varsigma})}{t_{k+\varsigma} - t_{k-\varsigma}}, \quad \begin{array}{l} \text{for } k = 1, \ldots T_m, \quad d = 1, \ldots, D, \\ m = 1, \ldots, M. \end{array}$$

$$(5.7)$$

The introduced delay of ς sampling periods smoothes the velocity data, in order to prevent the next key points to be selected too close to the previous key points. The velocities of the first and the last ς time instants are assigned to the constant vectors $v_m(t_{\varsigma+1})$ and $v_m(t_{T_{m-\varsigma}})$, respectively.

Table 5.1 LBG algorithm.

The input data is a set U consisting of T multidimensional vectors $\boldsymbol{\alpha}(t_k) \in \mathbb{R}^{2D}$, for $k = 1, 2, \ldots, T$.

1. **Initialization**: The number of partitions is set to $\rho = 1$, and the centroid $\bar{c}_1$ is set equal to the mean of the data vectors set $\{\boldsymbol{\alpha}(t_1), \boldsymbol{\alpha}(t_2), \ldots, \boldsymbol{\alpha}(t_T)\}$.

2. **Splitting**: Each centroid $\{\bar{c}_j + \delta\}$ is replaced by $\bar{c}_j + \delta$ and $\bar{c}_j - \delta$, where δ is a fixed perturbation vector.

3. **Assignment**: Each vector $\boldsymbol{\alpha}(t_k)$ is assigned to the nearest centroid, generating a set of clusters $\{Q_j\}_{j=1}^{\rho}$. The Euclidean distance from each data point to the cluster centroids is used as assignment criterion.

4. **Centroids updating**: The centroids are set equal to the mean of all data vectors assigned to each cluster, that is, $\bar{c}_j = \text{mean}\left(\boldsymbol{\alpha}(t_k), \text{for} \forall \boldsymbol{\alpha}(t_k) \in Q_j\right)$, for $j = 1, 2, \ldots, \rho$.

5. **Termination**: If the Euclidean distance between the data points and the assigned centroids is less than a given threshold, then stop. Otherwise, set $\rho = 2\rho$ and return to step 2.

The selection of the values of the parameters (e.g., ς) and other implementation details are discussed in Section 5.2.2.4, which is dedicated to the evaluation of the introduced approaches.

To avoid biased results caused by the different scaling in the numerical values of the positions, orientations, and velocities, the input data are first normalized to sequences with zero mean and unit variance for each dimension of the trajectories. The normalized poses and velocity vectors are denoted by $\hat{\mathbf{x}}_m(t_k)$ and $\hat{\mathbf{v}}_m(t_k)$, respectively. Then, these vectors are vertically concatenated into a combined two-dimensional (2D) vector, $\boldsymbol{\alpha}_m(t_k) = \left[\hat{\mathbf{x}}_m(t_k)^T \hat{\mathbf{v}}_m(t_k)^T\right]^T$. Afterward, the vectors from all trajectories are horizontally concatenated into a set of vectors U with the length equal to the total number of measurements from all trajectories, that is, $\sum_{m=1}^{M} T_m$. Next, the LBG classification algorithm is run on the set U to group all $\boldsymbol{\alpha}$ vectors into ρ clusters, by assigning each vector $\boldsymbol{\alpha}$ to the label of the closest centroid.

The candidate key points for the trajectories are assigned at the transitions between the cluster labels. The resulting sequences of key points after the initial segmentation with LBG algorithm are denoted by $\mathbf{K}_m^{\text{LBG}} = \left\{\kappa_{i,m}^{\text{LBG}} = (t_{i,m}, \mathbf{x}_{i,m})\right\}$, where the ith key point of the demonstration m (i.e., $\kappa_{i,m}^{\text{LBG}}$) is represented as an ordered set consisting of the time index t_k and the corresponding pose $\mathbf{x}_m(t_k)$.

In the presented implementation of the LBG algorithm, the total number of clusters ρ was set equal to an empirically selected value.

In addition, the number of iterations between steps 3 and 5 in Table 5.1 has been limited to 10, except for the final ρth partition in which complete convergence is allowed. The rationale for limiting the iterations is that there is no point of seeking full convergence of the centroids in each step, when they will be split and updated anyway. The two modifications improve the computational efficiency of the LBG algorithm.

To illustrate the LBG segmentation, an example of assigning key points for simple 2D trajectory with rectangular shape is shown in Figure 5.1 (the trajectory corresponds to painting the contour of a rectangular panel). The number of clusters ρ was empirically set to 16. Figure 5.1a presents the results when only the normalized position coordinates x and y are used for clustering. The key points, represented by circles in the figure, are associated with the changes in x and/or y position coordinates. The selected set of key points misses the corners, which represent important features of the demonstration, since they convey information about the spatial constraints of the task. Labeling with the LBG algorithm for the same trajectory based solely on the velocity features is shown in Figure 5.1b. In this case, there are significant velocity changes around the corners of the trajectories (due to slowing down of the motions around the corners). Figure 5.1c displays the LBG clustering results for combined positions and velocities data. When compared to the results in Figure 5.1a and b that use only positions or velocities features, the segmentation of the trajectory with combined positions and velocities in Figure 5.1c yields improved results, since it encapsulated the relevant trajectory features.

With regard to the number of clusters for partitioning a dataset, several works in the literature proposed methods for automatic selection of the optimal number of clusters. For instance, Davies and Bouldin (1979) formulated a coefficient of separation of the clusters based on the distance between the centroids of any two clusters, normalized by the standard deviation of the data within each cluster and by the number of data points assigned to each cluster. Similar measure of clusters' separation is the Silhouette coefficient (Kaufman and Rousseeuw, 1990), which calculates the average distance of a data point in a cluster to the other points in the same cluster, and compares it with the averaged distance of the point to the data points assigned to other clusters. A partition coefficient has been introduced in Bezdek (1981), based on soft (fuzzy) assignment of a data point to different clusters, followed by maximizing the data membership to the cluster. The aforementioned model validation coefficients are plotted as a function of the number of clusters, and subsequently the plots are used for selection of an optimal number of clusters. In addition, a body of works has investigated this problem from the aspect of probabilistic fitting of a cluster model to data. In this group of methods, commonly used criteria for estimation of the model size include Bayesian information criterion

(a)
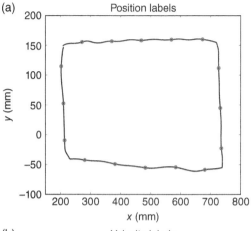

(b)
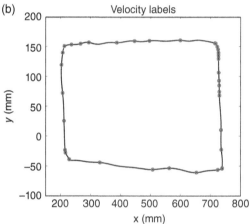

(c)
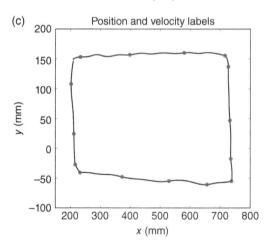

Figure 5.1 An example of initial selection of trajectory key points with the LBG algorithm. The key points are indicated using circles. The following input features are used for clustering: (a) normalized positions coordinates $(\hat{x}, \hat{y})$; (b) normalized velocities $(\hat{v}_x, \hat{v}_y)$; and (c) normalized positions and velocities $(\hat{x}, \hat{y}, \hat{v}_x, \hat{v}_y)$.

(Schwarz, 1978), minimum description length criterion (Rissanen, 1978), Akaike information criterion (Akaike, 1974), Bayes factors (Kass and Raftery, 1995), and cross-validated likelihood (Smyth, 1998). The criteria employ Bayesian formalism for finding a trade-off between maximizing the likelihood of a model in fitting the data and minimizing the number of clusters. Variants of these criteria have also been reported within the published literature (Berkhin, 2001).

The LBG algorithm can also automatically determine the number of clusters ρ, by setting a threshold distance value below which centroid splitting would cease. Although in this case the number of clusters does not need to be prespecified by the end-user, it requires specifying the threshold value for the stopping criterion. In addition, the resulting key points labeling could suffer from overfitting or missing important trajectory features.

For generic trajectories, determining the number of key points that is sufficient to represent the demonstrated trajectories represents a challenging problem. In cases where the number of identified key points is low, some relevant features of the trajectories may be excluded from the generalization process. On the other hand, extraction of too many key points may result in overfitting of the demonstrated trajectories.

Criteria for initial selection of key points have been reported in several works in the literature. Calinon and Billard (2004) suggested assigning a key point if there is a change in the direction for one of the position coordinates greater than a given threshold, or if the distance to the previous key point is greater than a given threshold. Asfour et al. (2006) proposed an extended set of criteria for the key points selection, which included the following: (i) the angle between two position vectors of the trajectory is less than a given threshold, (ii) the norm between two position vectors is greater than a given threshold, and (iii) the number of time frames between the previous key point and the current position is greater than a given threshold. This set of criteria is more complete, since key points can be identified even if there is no significant change in any of the individual coordinates, but there is an overall change in the direction of the demonstrated motions. The drawback of these two techniques for key points selection is the requirement for manual adjustment of the threshold parameters for different trajectories. For example, in the work of Asfour et al. (2006) the recorded trajectories consisted of 6D positions and orientations trajectories of the hand trajectories and 7D joint angles trajectories. Therefore, the process of selection of candidate key points required experimentally to determine the values of 15 threshold parameters.

An advantage of the method described in the book in comparison to the approaches described earlier is in assigning trajectory key points by using only the number of clusters ρ as a parameter, thus avoiding the manual tuning of a large set of parameters. The demonstrator leverages the learning process by prespecifying the number of clusters based on his/her prior knowledge of the geometry of the trajectories. The selected value of the parameter ρ can serve for segmentation purposes with a family of trajectories with similar level of complexity.

5.2.2 HMM-Based Modeling and Generalization

5.2.2.1 Related Work

HMM has been one of the most widely used methods for modeling and analysis of human motions, due to its robustness to spatiotemporal variations within sequential data (Pook and Ballard, 1993; Tso and Liu, 1997, Yang *et al.*, 1997). On the other hand, the problem of reproduction of human motions using HMMs for task representation has been less studied in the literature. Therefore, the research presented in the book concentrates on the generation of a generalized trajectory for reproduction of HMM modeled trajectories (Vakanski *et al.*, 2012).

Among the works in the literature that tackle the trajectory reconstruction problem, the approach reported by Tso and Liu (1997) used the observation sequence with the maximum likelihood of being generated by the trained HMM to be reproduced by the robot. However, because of the natural noise content of human movements, the demonstrated trajectories are generally not suitable for execution by the robot learner. In the work of Inamura *et al.* (2006), a large number of hidden state sequences for a learned HMM were synthetically generated, and a generalized trajectory was obtained by averaging over the set of trajectories that corresponded to the output distributions of the model for the given state sequences. A similar averaging approach was also employed in the article by Lee and Nakamura (2006). The drawback of the averaging approaches is that the smoothness of the reconstructed generalized trajectory cannot be guaranteed.

Another body of work proposed to generate a trajectory for reproduction of HMM modeled skills based on interpolation of trajectory key points (Inamura *et al.*, 2003). For example, Calinon and Billard (2004) first used the forward algorithm to find the observation sequence with the maximum likelihood of the trained HMM, and then the most probable sequence of hidden states for that observation sequence was obtained by applying the Viterbi algorithm. To retrieve a task reproduction trajectory, the extracted key points were interpolated by a third-order spline for the Cartesian trajectories and via a cosine fit for the joint angle trajectories. Although this approach generated a smooth trajectory suitable for robot reproduction, the resulting trajectory corresponded to only one observed demonstration. What is preferred instead is a generalized trajectory that will encapsulate the observed information from the entire set of demonstrations. The HMM-based key points approach for trajectory learning was also employed by Asfour *et al.* (2006). The authors introduced the term of common key points (which refers to the key points that are found in all demonstrations). Thus, the generalized trajectory included the relevant features shared by the entire set of individual demonstrations. Their approach was implemented for emulation of generalized human motions by a dual-arm robot with the generalized trajectory obtained by linear interpolation through the common key points.

The approach presented in the book is similar to the works of Calinon and Billard (2004) and Asfour *et al.* (2006), in using key points for describing the important parts of the trajectories for reproduction. Unlike these two approaches, a clustering technique is used for initial identification of candidate key points prior to the HMM modeling. The main advantage of the presented approach is in employing the key points from all demonstrations for reconstruction of a trajectory for task reproduction. Contrary to the work of Asfour *et al.* (2006), the key points that are not repeated in all demonstrations are not eliminated from the generalization procedure. Instead, by introducing the concept of null key points, all the key points from the entire set of demonstrations are considered for reproduction. Furthermore, by assigning low weights for reproduction to the key points that are not found in most of the demonstrations, only the most consisted features across the set of demonstrations will be reproduced. A detailed explanation of the approach is given in the following sections.

5.2.2.2 Modeling

5.2.2.2.1 HMM Codebook Formation

In this study, the modeling of the observed trajectories is performed with discrete HMMs, which requires the recorded continuous trajectories to be mapped into a codebook of discrete values. The technique described in Section 5.2.1 is employed for vector quantization purposes, that is, the data pertaining to the positions and velocities of the recorded movements are normalized, clustered with the LBG algorithm, and associated with the closest cluster's centroid. Resultantly, each measurement of the object's pose $\mathbf{x}_m(t_k)$ is mapped into a discrete symbol $o_{m,k}$ from a finite codebook of symbols $\{c_f\}_{f=1}^{N_o}$, that is, $o_{m,k} \in \{c_1, c_2, ..., c_{N_o}\}$, where N_o denotes the total number of observation symbols. The entire observation set thus consists of the observation sequences corresponding to all demonstrated trajectories, that is, $\{\mathcal{O}_m = (o_{m,1}, o_{m,2}, ..., o_{m,T_m})\}_{m=1}^{M}$, where M is the total number of demonstrations.

Several other types of strategies for codebook formation were attempted and compared to the adopted technique. Specifically, the multidimensional short-time Fourier transform (Yang *et al.*, 1997) has been implemented, which employs concatenated frequency spectra of each dimension of the data to form the prototype vectors, which are afterward clustered into a finite number of classes. Another vector quantization strategy proposed in Lee and Kim (1999) and Calinon and Billard (2004) was examined, which is based on homogenous (fixed) codebooks. This approach employs linear discretization in time of the observed data. Furthermore, a vector quantization based on the LBG clustering of the multidimensional acceleration, velocity, and pose data vectors was tested. The best labeling results were achieved by using the normalized velocity and position approach, which is probably due to the hidden dependency of the position and velocity features for human produced trajectories.

5.2.2.2.2 HMM Initialization

The efficiency of learning with HMMs in a general case depends directly on the number of available observations. Since in robot PbD the number of demonstrations is preferred to be kept low, a proper initialization of the HMM parameters is very important. Here, the initialization is based on the key points assignment with the LBG algorithm. The minimum distortion criterion (Linde *et al.*, 1980; Tso and Liu, 1997) is employed to select a demonstration X_σ, based on calculating the time-normalized sum-of-squares deviation, that is,

$$\sigma = \arg\min_{1 \le m \le M} \frac{\sum_{k=1}^{T_m} \left(\boldsymbol{\alpha}_m(t_k) - \bar{c}(o_{m,k}) \right)^2}{T_m}, \tag{5.8}$$

where $\boldsymbol{\alpha}_m(t_k)$ is the combined position/velocity vector of time frame k in the demonstration m, and $\bar{c}(o_{m,k})$ is the centroid corresponding to the label $(o_{m,k})$.

Accordingly, the number of hidden states N_s of HMM is set equal to the number of regions between the key points of the sequence X_σ (i.e., the number of key points of X_σ plus one). The HMM configuration known as a *Bakis left-right topology* (Rabiner, 1989) is employed to encode the demonstrated trajectories. The self-transition and forward-transition probabilities in the state-transition matrix are initialized as follows:

$$\begin{cases} a_{i,i} = \dfrac{1}{W} \left(1 - \dfrac{1}{\tau_{i,\sigma}} \right) \\[2ex] a_{i,i+1} = \dfrac{1}{W} \dfrac{1}{\tau_{i,\sigma}} \\[2ex] a_{i,i+2} = \dfrac{1}{W} \dfrac{1}{4\tau_{i,\sigma}} \end{cases}, \tag{5.9}$$

where $\tau_{i,\sigma}$ is the duration in time steps of the state i in demonstration X_σ, and W is a stochastic normalizing constant ensuring that $\sum_j a_{i,j} = 1$. The transitions $a_{i,i+2}$ allow states to be skipped in the case when key points are missing from some of the observed sequences. All other states transition probabilities are assigned to zeros, making transitions to those states impossible. The output probabilities are assigned according to

$$b_i(f) = \frac{n_{i,\sigma}(c_f)}{\tau_{i,\sigma}}, \tag{5.10}$$

where $n_{i,\sigma}(c_f)$ represents the number of times the symbol c_f is observed in state i for the observation sequence $\mathcal{O}_\sigma$. The initial state probabilities are set equal to $\boldsymbol{\pi} = \begin{bmatrix} 1 & 0 & \dots & 0 \end{bmatrix}$.

5.2.2.2.3 HMM Training and Trajectory Segmentation

The initialized values for the model parameters $\lambda = (\mathbf{A}, \mathbf{B}, \boldsymbol{\pi})$ are used for HMM training with the Baum–Welch algorithm (Section 4.2.3). All observed sequences $\mathcal{O}_1, \mathcal{O}_2, \dots, \mathcal{O}_M$ are employed for training purposes. Afterward, the Viterbi algorithm is used to find the most likely hidden states sequence $\mathcal{S}_m = s_{m,1}, s_{m,2}, \dots, s_{m,T_m}$ for a given sequence of observations $\mathcal{O}_m$ and a given model (Section 4.2.2). The transitions between the hidden states are assigned as key points. The starting point and the ending point of each trajectory are also designated as key points, because these points are not associated with the state transitions in the HMM. Also, due to the adopted Bakis topology of the HMM-state flow structure, certain hidden states (and key points as a consequence) may have not been found in all observed trajectories. Those key points are referred to as *null key points* in this work. Introducing the concept of null key points ensures that the length of the sequences of key points for all demonstrations is the same and equal to $N_s + 1$. Thus, the segmentation of the demonstrated trajectories with HMM results in a set of sequences of key points $K_m^{\text{HMM}} = \left\{ \kappa_{i,m}^{\text{HMM}} \right\}$, for $i = 1, 2, \dots, N_s + 1$, that correspond to the same hidden states (i.e., identical spatial positions) across all trajectories. Generalization of the demonstrations involves clustering and interpolation of these points, as explained in Section 5.2.2.3.

5.2.2.3 Generalization

5.2.2.3.1 Temporal Normalization of the Demonstrations via Dynamic Time Warping

Since the demonstrated trajectories differ in lengths and velocities, the extracted sequences of trajectory key points $\mathbf{K}_m^{\text{KMM}}$ correspond to different

temporal distributions across the trajectories. For generalization purposes, the set of key points needs to be aligned along a common time
vector. This will ensure tight parametric curve fitting in generating a
generalized trajectory through interpolation of the extracted key points.
Otherwise, the temporal differences of the key points across the demonstrations could cause spatial distortions in the retrieved reproduction
trajectory. The temporal reordering of the key points is performed with
the dynamic time warping (DTW) algorithm through the alignment of
the entire set of demonstrated trajectories with a reference trajectory.

To select a reference sequence for time warping of the demonstrations,
the forward algorithm (Section 4.2.1) is used for finding the log-likelihood
scores of the demonstrations with respect to the trained HMM, that is,
$\mathcal{L}(m) = \log P(\mathcal{O}_m | \lambda)$. The observation $\mathcal{O}_\varphi$ with the highest log likelihood
is considered as the most consistent with the given model; hence, the
respective trajectory X_φ is selected as a reference sequence.

Next, the time vector of the trajectory X_φ, that is, $\mathbf{t}_\varphi = \{t_1, t_2, ..., t_{T_\varphi}\}$, is
modified in order to represent the generalized time flow of all demonstra-
tions. This is achieved by calculating the average durations of the hidden
states from all demonstrations $\bar{\tau}_i$, for $i = 1, 2, ..., N_s$. Since the key points are
taken to be the transitions between the hidden states in the HMM, the
new key point time stamps for the demonstration $\mathcal{O}_\varphi$ are reassigned as
follows:

$$t_{k_j} = 1 + \sum_{i=1}^{j-1} \bar{\tau}_i, \text{ for } j = 2, ..., N_s + 1, \tag{5.11}$$

where t_{k_j} denotes the new time index of the jth key point, and $t_{k_1} = 1$ since
the first key point is always taken at the starting location of each trajectory.
Subsequently, a linear interpolation between the new time indexes of the
key points t_{k_j} is applied to find a warped time vector $\mathbf{t}_\varphi^{\text{warped}}$. The reference
sequence $X_\varphi^{\text{warped}}$ is obtained from the most consistent demonstrated
trajectory X_φ with time vector $\mathbf{t}_\varphi^{\text{warped}}$.

Afterwards, multidimensional DTWs are run on all demonstrated tra-
jectories against the time-warped data sequence $X_\varphi^{\text{warped}}$. It resulted in a set
of warped time waves, $G_1, G_2, ..., G_M$, that aligned every trajectory with the
$X_\varphi^{\text{warped}}$ trajectory. Accordingly, all key points are shifted in time by
reassigning the time index of every key point $\kappa_{j,m}$ with the time index
contained in the jth member of G_M. The resulting warped sequences
of key points are denoted by $K_m^{\text{DTW}} = \{\kappa_{j,m}^{\text{DTW}}\}$, for $j = 1, 2, ..., N_s + 1$.

With the key points from all demonstrations temporally aligned,
parametric curve fitting across each dimension would suffice to create

a generalized trajectory that represents the set of demonstrated trajectories. However, this approach would fail to separate the deliberate and precise motions from the transitional portions of the trajectories. Therefore, weighting coefficients are assigned for the different portions of trajectories, as bias coefficients of their relative importance for reproduction.

5.2.2.3.2 Key Points Weighting

Introducing the weighting coefficients was inspired by the works of Billard *et al.* (2004) and Calinon and Billard (2007), where the relative importance of each dimension from a set of measured variables was compared, and for reproduction purposes greater weights were imposed on the dimensions that were less variant across the different demonstrations.

Similarly, in this work weighting coefficients are assigned to all clusters of key points. The weighting is based on the assumption that the key points that are not closely clustered in comparison to the rest of the key points are generally less significant for reproduction. The weighting concept is illustrated in Figure 5.2.

The root mean-square (RMS) error is employed to measure the closeness of each cluster of key points as follows:

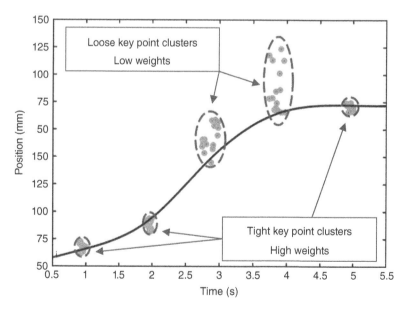

Figure 5.2 Illustration of the weighted curve fitting. For the clusters with small variance of the key points high weights for spline fitting are assigned, whereas for the clusters with high variance of the key points low weights are assigned, which results in loose fitting.

$$\vartheta(Q_j) = \sqrt{\sum_{m=1}^{M} \left(\kappa_{j,m}^{\mathrm{DTW}} - \bar{c}_j^{\mathrm{DTW}}\right)^2},$$ (5.12)

where $\bar{c}_j^{\mathrm{DTW}}$ denotes the centroid of the cluster Q_j of key points with time index j. For each cluster of key points Q_j, a normalized weighting coefficient is assigned based on

$$a_j = \begin{cases} 0 & ,\text{for} \quad \vartheta(Q_j) \geq \varepsilon_{\max} \\ \dfrac{\varepsilon_{\max} - \vartheta(Q_j)}{\varepsilon_{\max} - \varepsilon_{\min}} & ,\text{for} \quad \varepsilon_{\min} < \vartheta(Q_j) < \varepsilon_{\max}, \\ 1 & ,\text{for} \quad \vartheta(Q_j) \leq \varepsilon_{\min} \end{cases}$$ (5.13)

where $\varepsilon_{\min}$ and $\varepsilon_{\max}$ are empirically chosen threshold RMS error values. The introduced weighting scheme in (5.13) sets the weights to 0 and 1, when the RMS error is below or above a predefined threshold value, respectively. The upper threshold value $\varepsilon_{\max}$ is typically chosen at two or three standard deviations of the data points from the centroids of the corresponding cluster. The lower threshold value $\varepsilon_{\min}$ is typically set equal to a half of the standard deviation of the data points, in order to obtain high weights for reproduction of the data points that are closely clustered. The selection of the threshold values $\varepsilon_{\min}$ and $\varepsilon_{\max}$ for the particular trajectories used for evaluation of the presented method is explained in Section 5.2.2.4.

It should be noted that the null key points are not used for weighting or interpolation. Instead, they were introduced to ensure that the key points with the same index correspond to identical spatial positions across the trajectories.

5.2.2.3.3 Spline Fitting and Interpolation

After determining the weights of the clusters of key points, a generalized trajectory can be created by using a number of curve-fitting techniques. A penalized cubic smoothing-spline regression (Rice and Rosenblatt, 1983) is used here for that purpose. This technique is well suited for fitting of a smooth curve to scattered data, such as generation of a smooth trajectory from a set of key points from multiple trajectories. Each dimension of the trajectories is fitted parametrically over time with the corresponding weighting coefficients (see Figure 5.2). The spline curve is interpolated at intervals equal to the sampling period, resulting in a generalized trajectory X_{gen} suitable for following by the learning robot.

5.2.2.4 Experiments

A block diagram that describes the high-level information flow for the developed approach is presented in Figure 5.3. The algorithm automatically produces a generalized version of the set of demonstrated trajectories, based on a series of data processing steps. A human operator intervened in collecting the demonstrated trajectories, and in the transfer of the derived strategy for task reproduction to the robotic platform for execution.

To evaluate the approach, two experiments are conducted for a manufacturing process of painting. A hand tool is moved by an operator in a simulated painting process, by traversing the areas of the parts that are required to be painted. The first experiment follows a simple trajectory, while the second experiment entails a more complex geometry and trajectory that includes nonplanar regions and waving motions of the painting tool with different amplitudes.

The experiments were conducted in the National Research Council (NRC)—Aerospace Manufacturing Technology Centre (AMTC) in Montréal, Canada.

5.2.2.4.1 Experiment 1

The experimental setup for the first experiment is shown in Figure 5.4. A demonstrator is performing virtual painting of a panel using a hand tool that serves as a spray gun. The attached optical markers on the tool are tracked by the optical tracking system Optotrak (Section 2.1) shown in Figure 5.4b. The task trajectory consists of (i) approaching the upper left-side corner of the panel from an initial position of the tool, (ii) painting the rectangular contour of the panel in a clockwise direction, (iii) painting the inner part of the panel with waving left-to-right motions, and (iv) bringing the tool back to the initial position.

The task is demonstrated four times by four different operators, that is, the measurement set consists of trajectories X_m, for $m = 1, 2, ..., 16$, containing the position and orientation data of the tool with respect to a reference frame (Figure 5.4c). The robustness of the learning process can be enhanced by eliminating the demonstrations that are inconsistent with the acquired dataset (Aleotti and Caselli, 2006). Otherwise, the learning system might fail to generalize correctly from a set of demonstrations which are too dissimilar. The sources of variations can be attributed to the uncertainties arising from (i) involuntary motions of the demonstrator, (ii) different conditions in performing the demonstrations, (iii) difference in the performance among the demonstrators, (iv) fatigue due to multiple demonstrations of the same task, etc. The presented work emphasizes on the problem of inconsistency within a set of demonstrations executed by different human demonstrators, through the introduction of a preliminary step for evaluation of the demonstrators' performance.

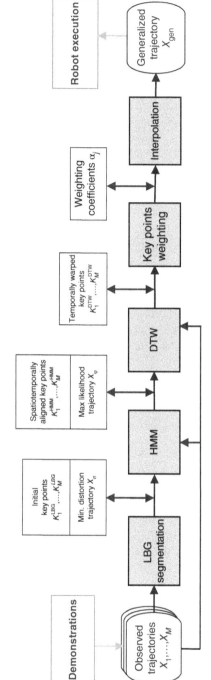

Figure 5.3 Diagram representation of the presented approach for robot PbD. The solid lines depict automatic steps in the data processing. For a set of observed trajectories $X_1, \ldots, X_M$, the algorithm automatically generates a generalized trajectory X_{gen}, which is transferred to a robot for task reproduction.

(a)

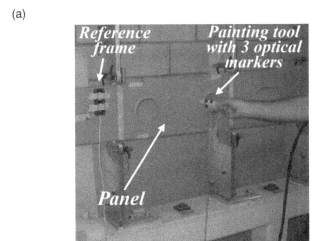

(b)

(c)

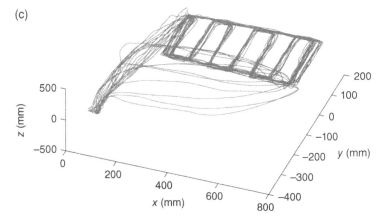

Figure 5.4 (a) Experimental setup for Experiment 1: panel, painting tool, and reference frame; (b) perception of the demonstrations with the optical tracking system; and (c) the set of demonstrated trajectories.

This step carries out an initial comparison of the distributions of the first, second, and third derivative of position with respect to time, that is, velocities, accelerations, and jerks, respectively, of the demonstrated trajectories. For each subject, the values of the parameters from all trajectories are combined, and the distributions of the parameters are represented with box plots in Figure 5.5. The plots, for example, indicate that the demonstrations by Subject 4 have the greatest deviations of the accelerations and the jerks, while for the rest of the subjects the distributions are more uniform. Therefore, the demonstrations by Subject 4 were eliminated from the recorded dataset. These parameters can also be exploited as criteria for evaluation of the demonstrations with regard to the motor capabilities of the robot learner. For instance, if the levels of accelerations of the demonstrations are greater than the maximum accelerations available from the robot links motors, the executed task will differ from the demonstrated task, and it can even lead to a task failure, depending on the circumstances. Consequently, the input set of trajectories is reduced to $M = 12$ trajectories, demonstrated by Subjects 1, 2, and 3. This initial refinement results in a more consistent initial dataset, and it leads to generation of a more robust task model. The Cartesian positions of the demonstrations are shown superimposed in Figure 5.4c. The number of measurements of the demonstrations (i.e., T_m) varies from 3728 to 7906 measurements, with an average of 5519. One can infer that despite the similarity of the trajectories' geometry in Figure 5.4c, the velocities of the demonstrations varied significantly between the subjects.

After an initial preprocessing of the data, which included smoothing of the recorded trajectories with a Gaussian filter of length 10, candidate key points are identified using the LBG algorithm, as explained in Section 5.2.1. The parameter ς, related to the delay of time frames for the velocity calculation, is chosen equal to 20 sampling periods (i.e., 200 milliseconds), and the number of clusters ρ is set to 64. These parameters are chosen heuristically to balance the complexity and the duration of the demonstrated trajectories. The number of key points per trajectory ranges from 77 to 111 (average value of 87.75 and standard deviation of 11.7). For the minimum distortion trajectory X_{12} (found using (5.8)), 82 key points are identified. The key points which are closer than 20 time frames are eliminated from the minimum distortion trajectory, resulting in 72 key points in total (shown in Figure 5.6).

Afterward, the recorded trajectories are mapped into discrete sequences of observed symbols, and discrete HMMs are trained. For that purpose, Kevin Murphy's HMM toolbox (Murphy, 1998) is used in MATLAB environment. The number of discrete observation symbols N_o is set to 256. The number of hidden states N_s is set equal to the number of key points of the minimum distortion trajectory plus 1, that is, 73. The Viterbi

Figure 5.5 Distributions of (a) velocities, (b) accelerations, and (c) jerks of the demonstrated trajectories by the four subjects. The bottom and top lines of the boxes plot the 25th and 75th percentile for the distributions, the bands in the middle represent the medians, and the whiskers display the minimum and maximum of the data.

(a)
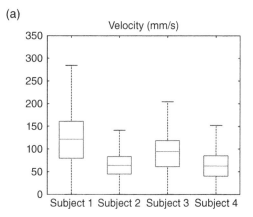

(b)
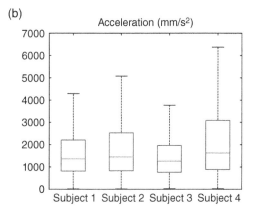

(c)
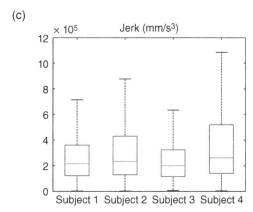

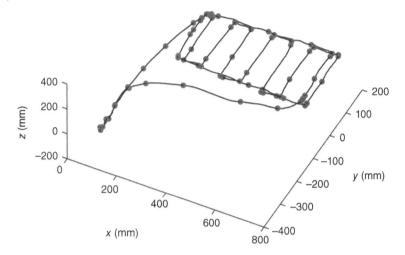

Figure 5.6 Initial assignment of key points for the trajectory with minimum distortion.

algorithm is employed for finding the sequences of hidden states for each observation sequence, and key points are assigned to the position and orientation values of the trajectories that correspond to the transitions between the hidden states in the observation sequences. Thus, including the starting and ending points of the trajectories, each demonstration was described by a time-indexed sequence of 74 key points. As explained earlier, for the observation sequences with missing hidden states, the coordinates of the corresponding missing key points are set to zero (for that reason, they were referred to as null key points). This procedure results in spatiotemporal ordering of the key points for all trajectories, with a one-to-one mapping existing between the key points across the different demonstrations (e.g., for each trajectory, the key point with the index $j = 16$ corresponds to the upper right corner of the contour). Log likelihoods of the observation sequences for the trained HMM are obtained with the forward algorithm. The observation sequence $\mathcal{O}_3$ had the highest log likelihood; therefore, the demonstration X_3 is used for temporal alignment of the set of trajectories with the DTW algorithm.

To obtain a generalized trajectory that would represent the entire set, its length should be approximately equal to the average length of the trajectories in the set. Hence, the average durations of the hidden states for the trained HMM are calculated, whereby the newly created sequence of hidden states has length of 5522 time frames. This value is indeed very close to the average length of the demonstrations, being 5519. This time vector is used to modify the time flow of the demonstration X_3 as explained in Section 5.2.2.3.1. Subsequently, the resulting sequence is used as a

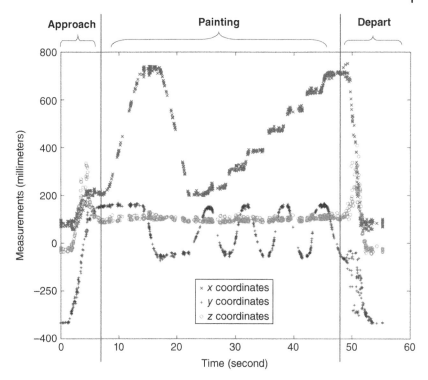

Figure 5.7 Spatiotemporally aligned key points from all demonstrations. For the parts of the demonstrations which correspond to approaching and departing of the tool with respect to the panel, the clusters of key points are more scattered, when compared to the painting part of the demonstrations.

reference sequence for the DTW algorithm, that is, to align each of the demonstrated trajectories against it. The DTW alignment resulted in time-warped trajectories with length of 5522 time frames, from which the time instances of the key points are extracted. The Cartesian positions coordinates x, y, and z of the DTW-reassigned key points from all demonstrations are shown in Figure 5.7.

For the approaching and departing parts of trajectories in Figure 5.7, the variance among the key points is high, which implies that these parts are not very important for reproduction. The distributions of the RMS errors for the clusters of key points along x, y, and z coordinates are shown in Figure 5.8a. There are 73 clusters of key points in total, where the starting and ending subsets of key points, which correspond to the approaching and departing parts of the trajectories, exhibit larger deviations from the centroids of the clusters. Therefore, weighting coefficients with lower values are assigned for those parts of the trajectories. The variance of the

(a)

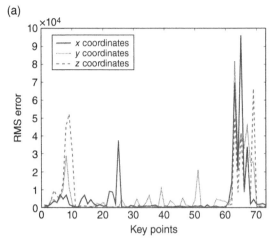

Figure 5.8 (a) RMS errors for the clusters of key points, (b) weighting coefficients with threshold values of 1/2 and 2 standard deviations, and (c) weighting coefficients with threshold values of 1/6 and 6 standard deviations.

(b)

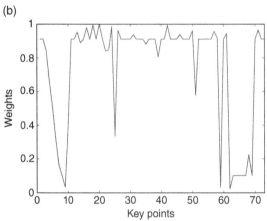

(c)

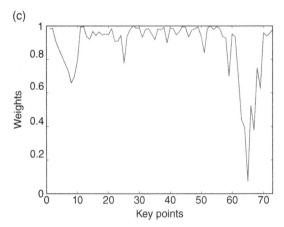

key points associated with painting the contour and the inner side of the panel is low, and thus the computed weighting coefficients have values close to 1. Values of 0.5 and 2 standard deviations of the RMS errors for the corresponding cluster are adopted for $\varepsilon_{\min}$ and $\varepsilon_{\max}$, respectively, in (5.13). The weights for such adopted thresholds are displayed in Figure 5.8b. For comparison, the weighting coefficients for the key point clusters with the threshold values of 1/6 and 6 standard deviations are shown in Figure 5.8c. In this case, the weights distribution will result in a reproduction trajectory that will fit more closely the demonstrated trajectories. However, such choice of the parameters will not have very significant impact on the final outcome, that is, the approach is not highly sensitive to the choice of the threshold values $\varepsilon_{\min}$ and $\varepsilon_{\max}$.

The last step entails smoothing spline interpolation through the key points by using an empirically determined smoothing factor $f_s = 0.975$, which gives both satisfactory smoothness and precision in position for the considered application. For the orientation of the tool expressed in Euler roll–pitch–yaw angles, the 1D generalization results are given in Figure 5.9. The plots show the angle values of the key points and the interpolated curves for each of the angle coordinates. A plot of the Cartesian coordinates of the resulting generalized trajectory X_{gen} is shown in Figure 5.10.

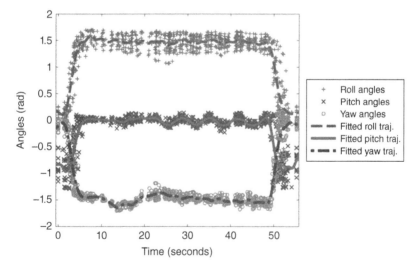

Figure 5.9 Generalization of the tool orientation from the spatiotemporally aligned key points. Roll angles are represented by a dashed line, pitch angles are represented by a solid line, and yaw angles are represented by a dash–dotted line. The dots in the plot represent the orientation angles of the key points.

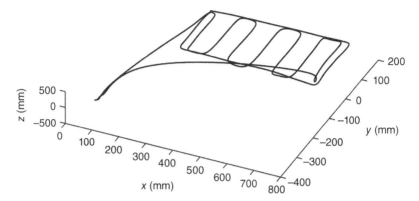

Figure 5.10 Generalized trajectory for the Cartesian *x–y–z* position coordinates of the object.

For comparison, the distributions of velocities, accelerations, and jerks for the demonstrated trajectories by the three subjects and the generalized trajectory are shown in Figure 5.11. First, from the plot of velocity distributions, it can be concluded that the mean value of the generalized trajectory is approximately equal to the average value of the velocities from the demonstrations by all subjects. This is the result of the generalization procedure, that is, the demonstrations by all subjects are used to find a generalized trajectory that will represent the entire set of demonstrations. Second, the distributions of accelerations indicate that the values of the generalized trajectory are lower than the values of the demonstrated trajectories. It is due to the fact that spline smoothing interpolation is employed for reconstruction of the generalized trajectory, which reduces the changes of velocities present in human motions. Third, the jerk distribution for the generalized trajectory is significantly condensed when compared to the demonstrated trajectories. Namely, the inconsistency of the human motions represented by high levels of jerk is filtered out by the generalization procedure, resulting in a smooth trajectory with low levels of jerk.

5.2.2.4.2 Experiment 2

The second experiment entails painting a panel with more complex geometry, shown in Figure 5.12a. The objective is to paint the top side of the panel, and the horizontal plane on the right-hand side of the panel. Both areas are shown bordered with solid lines in Figure 5.12a. The three optical markers shown in the picture are used to define the reference coordinate system of the panel. The task is demonstrated five times by

Figure 5.11 Distributions of (a) velocities, (b) accelerations, and (c) jerks for the demonstrated trajectories by the subjects and the generalized trajectory.

(a)

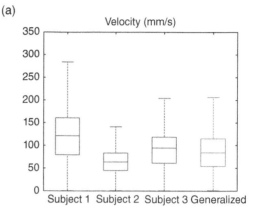

(b)

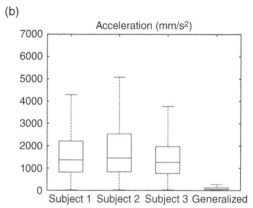

(c)

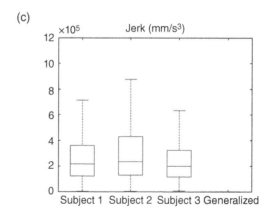

(a)

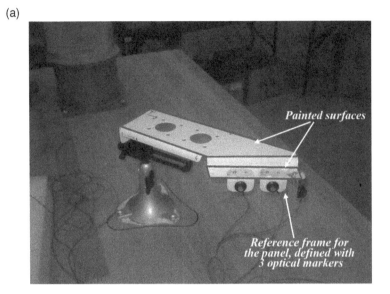

(b)

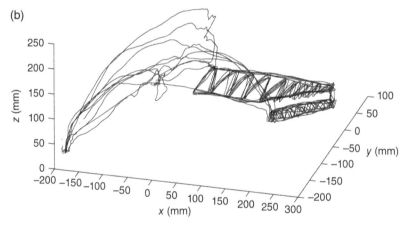

Figure 5.12 (a) The part used for Experiment 2, with the surfaces to be painted bordered with solid lines and (b) the set of demonstrated trajectories.

a single demonstrator, and all trajectories are shown superimposed in Figure 5.12b. Different from Experiment 1 where the shape of the generalized trajectory for the painting part is obvious from Figure 5.4c, in the case of Experiment 2 it is not easy to infer the shape of the generalized trajectory, especially for the small right-hand plane where the waving motions are not controlled.

The lengths of the demonstrations range between 4028 and 4266 measurements. The initial selection of candidate key points with the number of clusters ρ set to 64 results in numbers of key points between 122 and 134 per trajectory. Based on the sequence with minimum distortion, the number of hidden states N_s for training the HMM is set to 123. Afterward, the demonstrations are temporally aligned with DTW, with the length of the sequence of average states duration equal to 4141 time frames. The resulting set of key points from all five demonstrations is interpolated to generate a generalized trajectory X_{gen} (Figure 5.13a). The plot also shows the approach and retreat of the painting tool with respect to the panel.

The generalized trajectory for the pose of the painting tool was transferred to a desktop robot (Motoman$^\circledR$ UPJ), and afterward it was executed

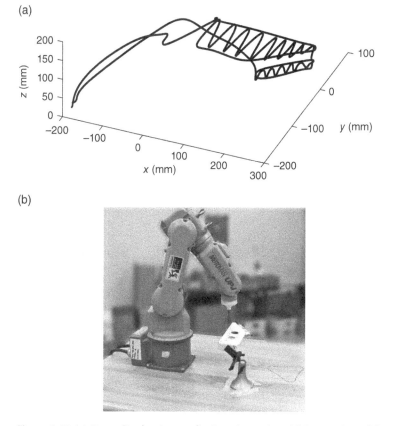

Figure 5.13 (a) Generalized trajectory for Experiment 2 and (b) execution of the trajectory by the robot learner.

by the robot (Figure 5.13b). The height of the robot's end-effector was set up to 30 millimeters above the part. The objective of the experiments was not to perform the actual painting with the robot, but to illustrate the approach for learning complex trajectories from observations of demonstrated tasks. The results demonstrate that with encoding of demonstrations at the trajectory level, complex demonstrated movements can be modeled and generalized for reproduction by a robot in a PbD environment.

5.2.2.5 Comparison with Related Work

To compare the presented method with the similar works in the literature, the obtained trajectory for task reproduction based on the approach reported by Calinon and Billard (2004) is shown in Figure 5.14. The demonstration set for Experiment 1 from the Section 5.2.2.4.1 is used, and the result is obtained by using the same parameter values reported in the aforementioned section. The initial key points selection based on LBG algorithm is employed, in order to compare the approaches on the most similar basis. The trajectory for task reproduction is generated by smoothing spline interpolation of the key points generated from the Viterbi path

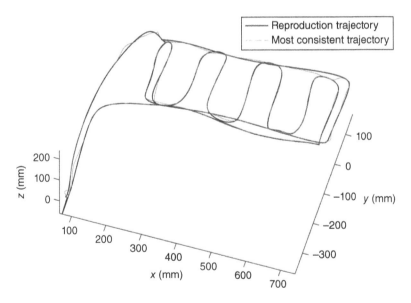

Figure 5.14 Generated trajectory for reproduction of the task of panel painting for Experiment 1, based on Calinon and Billard (2004). The most consistent trajectory (dashed line) corresponds to the observation sequence with the highest likelihood of being generated by the learned HMM.

of the most consistent trajectory, as explained in the work of Calinon and Billard (2004). The most consistent trajectory is shown with the dashed line in Figure 5.14. The main drawback of this approach is that it does not generalize from the demonstrated set. Instead, the reproduction trajectory is obtained by using a criterion for selecting the best demonstration from the recorded set, followed by a smoothing procedure. On the other hand, in the approach presented in the book the trajectory for the task reproduction (shown in Figure 5.10) is obtained by taking into account important features from the entire set of demonstrated trajectories, and it does not rely so heavily on only one of the demonstrated trajectories.

Another advantage of the presented approach with respect to the work by Calinon and Billard (2004) is in providing a mechanism for interpretation of the intent of the demonstrator. Based on the variance of the key points corresponding to the same spatial features across the demonstrated set, the method presented here assigns different weights for reproduction purposes. For instance, for the approaching and departing (transitional) sections of the trajectories, the variability across the demonstrated set is high, and subsequently the generalized trajectory does not follow strictly the trajectories key points. On the other hand, the section of the trajectories corresponding to painting the panel is more constrained, which is reflected in the generation of the generalized trajectory with tight interpolation of the identified key points. It can be observed in Figure 5.10 that the approaching and departing sections are smoother, due to the higher variance across the demonstrations, whereas in Figure 5.14 the approaching and departing sections simply follow the most consistent trajectory. The procedure adopted in the book has similarities to the work of Calinon (2009) based on GMM/GMR, where the sections of the demonstration with high values of the Gaussian covariances are considered less important for reproduction purposes, and vice versa.

The approach of Asfour *et al.* (2006), which employs generalization from the set of demonstrations by using the concept of common key points, is similar to the presented approach, although it may provide suboptimal results in some cases. For instance, if a key point is missing in one of the demonstrated trajectories, and it is present in all of the remaining trajectories, this key point will be eliminated as not being common and will not be considered for generalization purposes. To avoid this case, the approach might require applying a heuristic for analysis of the demonstrations, and elimination of the suboptimal demonstrations.

The results from the developed method and the methods reported in Calinon and Billard (2004) and Asfour *et al.* (2006) are compared by using the RMS error as a metric for evaluation of learned trajectories (Calinon et al., 2010)

$$\vartheta_{m_1,m_2} = \sum_k \left\| \mathbf{x}_{m_1}(t_k) - \mathbf{x}_{m_2}(t_k) \right\|, \quad \text{for } k = 1,2,\ldots, \tag{5.14}$$

where m_1 and m_2 are used for indexing the trajectories. The trajectories were scaled to sequences with the same length by using two techniques: linear scaling and DTW alignment. With the linear scaling technique, the length is set equal to the average length of the demonstrated trajectories. For the DTW scaling, the entire set of trajectories is aligned against one of the demonstrated trajectories. Rather than selecting the generalized trajectory for scaling the dataset with the DTW, some of the demonstrated trajectories were used as reference sequences for alignment, to avoid biased results for the RMS differences.

A color map chart with the RMS differences for the linearly scaled trajectories for Experiment 1 is shown in Figure 5.15. The first three trajectories in the chart represent the generalized trajectories obtained by the presented approach—X_{G1}, by the approach proposed in Calinon and Billard (2004)—X_{G2}, and the one proposed in Asfour et al. (2006)—X_{G3}.

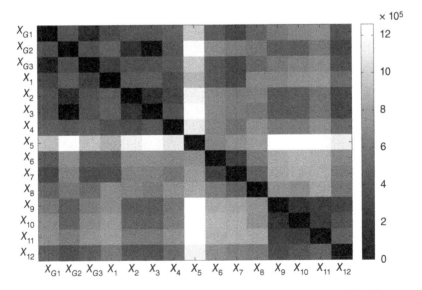

Figure 5.15 RMS differences for the reproduction trajectories generated by the presented approach (X_{G1}), the approaches proposed in Calinon and Billard (2004) (X_{G2}), Asfour et al. (2006) (X_{G3}), and the demonstrated trajectories (X_1–X_{12}). As the color bar on the right side indicates, lighter nuances of the cells depict greater RMS differences.

The rest of the cells correspond to the demonstrated trajectories X_1–X_{12}. The darker colors in the chart pertain to cells with smaller RMS differences. For instance, the trajectory X_5 is the most dissimilar to the whole set, and it can be noted that the trajectories demonstrated by the same subject X_1–X_4, X_5–X_8, and X_9–X_{12} are more similar when compared to the trajectories demonstrated by the other subjects.

The sums of the RMS differences for the trajectories from Figure 5.15 are given in Figure 5.16a. The RMS differences for scaling of the trajectories with DTW are also displayed. The demonstrated trajectories X_4, X_5, and X_{12} are chosen as reference sequences for aligning the trajectories. The DTW algorithm is repeated for three of the demonstrated trajectories

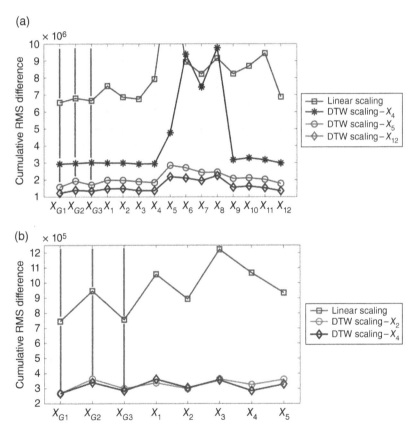

Figure 5.16 Cumulative sums of the RMS differences for the reproduction trajectories generated by the presented approach (X_{G1}), the approaches proposed in Calinon and Billard (2004) (X_{G2}), and Asfour et al. (2006) (X_{G3}). (a) Demonstrated trajectories (X_1–X_{12}) from Experiment 1 and (b) demonstrated trajectories (X_1–X_5) from Experiment 2.

to avoid any subjective results that can arise from using a particular trajectory as a reference sequence. As expected, the DTW algorithm aligns the trajectories better than the linear scaling technique. For the presented four cases, the cumulative RMS differences for the trajectory obtained by the presented approach (X_{G1}) are the smallest, which infers that it represents the best match for the demonstrated set.

Similarly, for Experiment 2 from the Section 5.2.2.4.2, the sums of the RMS differences are presented in Figure 5.16b. The DTW alignment is repeated for two of the demonstrated trajectories, X_2 and X_4. Based on the presented results, it can be concluded that the generalized trajectory generated by the described approach fits better the demonstrated set, when compared to the similar approaches proposed by Calinon and Billard (2004) and Asfour *et al.* (2006).

Another work in the literature that has some similarities to the one presented in the book has been reported by Gribovskaya and Billard (2008). In their study, DTW was employed initially to align the recorded sequences. Afterward, a continuous HMM was employed for encoding the trajectories. One drawback of such procedure is that the DTW algorithm distorts the trajectories. As a result, important spatial information from the demonstrations may be lost. In addition, the velocity profile of the DTW aligned demonstrations is deformed. In the method presented here, HMM is used for segmentation of the trajectories as recorded, and DTW is used before the interpolation procedure, only to shift the key points to a trajectory with a given time duration (which is found as the average of time durations of the hidden states form HMM). Hence, the velocity information is preserved for modeling the demonstrations with HMM, and the DTW affects the trajectories only at the level of key points.

The computational expense of the presented approach is compared with the state-of-the-art approach reported in Calinon (2009), which uses GMM and GMR for modeling and generalization of continuous movements in a robot PbD setting. The simulations are run on a 2.1 GHz dual-core CPU with 4 GB of RAM running under Windows XP in the MATLAB environment. Since there are slight variations for the processing times in different simulation trials, the codes were run five times, and the mean values and standard variations are reported. The computation times for obtaining a generalized trajectory from the sets of measured trajectories for both experiments from Section 5.2.2.4 are presented in Table 5.2. The DTW alignment step of the method is performed with a MATLAB executable (MEX) file, which increases the computational speed of the DTW algorithm for about 6–8 times compared to the standard MATLAB m-files. The results of the computational cost for the presented approach displayed in Table 5.2 indicate that most of the

Table 5.2 Mean values and standard deviations of the computation times for learning the trajectories from Experiment 1 and Experiment 2.

	CPU times (seconds)	
Steps of the program code:	Experiment 1	Experiment 2
1. Initial preprocessing (smoothing, removing NANs)	0.62 (±0.27)	0.27 (±0.01)
2. LBG clustering	38.38 (±23.20)	5.69 (±0.50)
3. HMM initialization	185.17 (±27.05)	51.33 (±16.76)
4. HMM training	108.41 (±22.36)	149.43 (±36.01)
5. HMM inference	24.40 (±0.81)	12.74 (±0.72)
6. DTW alignment	44.50 (±0.35)	23.92 (±0.37)
7. Key point weighting and interpolation	2.56 (±1.04)	2.28 (±0.01)
Total	**404.05 (±26.81)**	**245.73 (±39.21)**

Table 5.3 Mean values and standard deviations of the computation times for learning the trajectories from Experiment 1 and Experiment 2 by applying the GMM/GMR approach.

	CPU times (seconds)	
Steps of the program code:	First experiment	Second experiment
1. Initial preprocessing	0.49 (±0.01)	0.21 (±0.01)
2. DTW alignment	40.16 (±0.11)	20.03 (±0.38)
3. GMM encoding	847.77 (±140.27)	948.60 (±197.95)
4. GMR	0.70 (±0.10)	1.22 (±0.02)
Total	**889.13 (±140.17)**	**953.86 (±197.88)**

processing time is spent on the learning of HMM parameters and on the discretization of the continuous trajectories.

The computation times for the GMM/GMR method are reported in Table 5.3, with using the same MATLAB MEX subroutine for performing the DTW alignment phase. For Experiment 1, the trajectory X_{12} (with the length of 4939 measurements) is selected as a reference sequence for aligning the set of 12 demonstrated trajectories, whereas for Experiment 2 the trajectory X_2 (4048 measurements) is selected as a reference sequence for aligning the set of five trajectories. For the steps of GMM encoding and GMR generalization, the MATLAB codes provided in Calinon (2009) are used. The number of Gaussian components for modeling the trajectories was approximated by trial and error, and the values used for the reported times are 40 for the first experiment and 75 for the second experiment.

The generalized trajectories generated from the GMM/GMR method are given in Figure 5.17. Based on the results from Table 5.3, it can be concluded that the computational cost of modeling the 6D input data into a large number of Gaussians is computationally expensive. The total computation times with the presented approach are less than the times obtained with the GMM/GMR approach for both sets of trajectories.

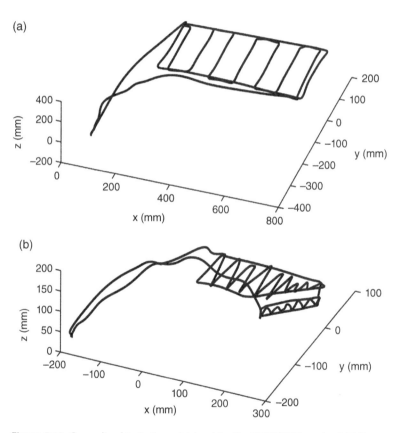

Figure 5.17 Generalized trajectory obtained by the GMM/GMR method (Calinon, 2009) for (a) Experiment 1 and (b) Experiment 2, in Section 4.4.4.

5.2.3 CRF Modeling and Generalization

5.2.3.1 Related Work

One of the application domains where CRFs have been most extensively utilized is the language processing. Examples include part-of-speech tagging (Lafferty *et al.*, 2001; Sutton and McCallum, 2006), shallow parsing (Sha and Pereira, 2003), and named-entity recognition (McDonald and Pereira, 2005; Klinger *et al.*, 2007). CRFs were reported to outperform HMMs for classification tasks in these studies. Other areas of implementation include image segmentation (He *et al.*, 2004; Quattoni *et al.*, 2005), gene prediction (DeCaprio *et al.*, 2007), activity recognition (Sminchisescu *et al.*, 2005; Vail *et al.*, 2007a), generation of objects' trajectories from video sequences (Ross *et al.*, 2006), etc.

In the robot PbD environment, Kjellstrom *et al.* (2008) presented an approach for grasp recognition from images using CRF. Martinez and Kragic (2008) employed the support vector machine (SVM) algorithm for activity recognition in robot PbD by modeling the demonstrated tasks at a symbolic level, followed by using CRF for temporal classification of each subtask into one of the several predefined action primitives classes. The comparison results of recognition rates between HMM and CRF indicated similar performance of both methods for short actions, and higher recognition rates by CRF for the case of prolonged continuous actions.

CRF has not been implemented before for acquisition of skills at the trajectory level in the area of robot PbD. Therefore, Section 5.2.3.2 presents one such approach (Vakanski *et al.*, 2010). Similarly to the approach presented in Section 5.2.2 that employed HMM for encoding trajectories, the hidden states of the demonstrator's intention in executing a task are related to the trajectory key points in the observed sequences. CRF is employed for conditional modeling of the sequences of hidden states based on the evidence from observed demonstrations. A generalized trajectory for reproducing the required skill by the robot is generated by cubic spline interpolation among the identified key points from the entire set of demonstrated trajectories.

5.2.3.2 Feature Functions Formation

The developed framework for robot learning from multiple demonstrations employs the optical motion capture device Optotrak Certus® (NDI, Waterloo, Canada) for the task perception (Section 2.1). Hence, the observation sequences are associated with the continuous trajectories captured during the demonstration phase. This section discusses the formation of the feature functions $\Theta(s_k, s_{k-1}, \mathcal{O}, t_k)$ in a linear chain CRF structure (Section 4.3.1), for the case when the observation sequences are human demonstrated trajectories. In general, CRFs can handle observed features of either continuous or discrete nature, whereas the hidden state variables are always of discrete nature. Structuring of the feature functions for both continuous and discrete observations is discussed next.

A CRF model designed for classification of human activities from continuous trajectories as observation data is presented in the work of Vail *et al.* (2007b). The authors introduced the following continuous-state feature functions:

$$\Gamma_{i,d}(s_k, \mathcal{O}, t_k) = I(s_k = i) o_k^{(d)},$$ (5.15)

for $i \in \{1, N_s\}$, where N_s denotes the number of hidden states, and d is used for indexing the dimensionality of the observed Cartesian positions o_k. The notation $I(s_k = 1)$ pertains to a binary indicator function, which equals to 1 when the state at time t_k is i, and 0 otherwise. To enhance the classification rate of the CRF in the article of Vail *et al.* (2007b), additional continuous-state feature functions are added to the model, corresponding to the observed velocities, squared positions, etc.

The transition feature functions for encoding the transition scores from state i at time t_{k-1} to state j at time t_k are defined as follows:

$$\Phi_{i,j}(s_k, s_{k-1}, \mathcal{O}, t_k) = I(s_{k-1} = i) I(s_k = j)$$ (5.16)

for $i, j \in \{1, N_s\}$.

An important remark reported in the article (Vail *et al.*, 2007b) is that the empirical sum of the features in the log-likelihood function (4.23) can be poorly scaled when continuous features functions are used. This can cause slow convergence of the optimization algorithms, or in the worst case, the algorithm would not converge. As a remedy, the authors proposed to perform normalization of the continuous observations to sequences with zero mean and unity variance, which improved the classification results.

Nevertheless, most of CRF's applications in the literature deal with categorical input features (e.g., labeling words in a text), rather than continuous measurements. In that case, the observed features are mapped to a finite set of discrete symbols $o_{m,k} \in \{c_1, c_2, \ldots, c_{N_o}\}$. The state feature functions are defined as binary indicators functions for each state-observation pairs as follows:

$$\Gamma_{i,f}(s_k, \mathcal{O}, t_k) = I(s_k = i) I(o_k = c_f)$$ (5.17)

for $i \in \{1, N_s\}$ and $f \in \{1, N_o\}$.

The transition feature functions $\Phi(s_k, s_{k-1}, \mathcal{O}, t_k)$ are defined in an identical manner as in (5.16). The structure of CRFs allows additional observed features to be easily added to the model by generalizing the feature functions. For instance, the dependence of state transitions to the observed symbol can be modeled by adding additional feature functions $I(s_k = i)I(s_{k-1} = j)I(o_k = c_f)$.

5.2.3.3 Trajectories Encoding and Generalization

The implementation of CRF in this work is based on discrete observation features, due to the aforementioned scalability problems with the continuous features. For that purpose, the continuous trajectories from the demonstrations are mapped to a discrete set of symbols by using the LBG algorithm, as described in Section 5.2.1. The output of the LBG clustering procedure is a set of discrete sequences $\mathcal{O}_1, \mathcal{O}_2, ..., \mathcal{O}_M$.

Training the CRF model requires provision of the corresponding sequences of hidden states (i.e., labels) $\mathcal{S}_m$ for each observation sequence $\mathcal{O}_m$. Similar to the approach described in Section 5.2.2, the labels are assigned to the key points along the trajectories, which are characterized by significant changes in position and velocity. The initial labeling of the trajectories is performed with the LBG technique. The clustering is carried over a set of concatenated normalized positions and velocities vectors from all demonstrated trajectories, that is, $\alpha_m(t_k) = \left[\hat{x}_m^{(d)}(t_k)^T \hat{v}_m^{(d)}(t_k)^T \right]^T$ for $k \in (1, T_m)$, $m \in (1, M)$, and $d \in (1, D)$. The transitions between the cluster labels are adopted as key points. The detected initial key points resulted in the labeled sequences denoted as $\mathcal{S}_{1,\text{init}}, \mathcal{S}_{2,\text{init}}, ..., \mathcal{S}_{M,\text{init}}$.

A common challenge associated with the task modeling in robot PbD is the limited number of demonstrations available for estimation of the model parameters (since it may be frustrating for a demonstrator to perform many demonstrations, and in addition, the quality of the performance can decrease due to fatigue or other factors). Therefore, to extract the maximum information from limited training data, as well as to avoid testing on the training data, the leave-one-out cross-validation technique is adopted for training purposes. This technique is based on using a single observation sequence for validation, and using the remaining observation sequences for training. For instance, to label the observed sequence which corresponds to the third demonstration $\mathcal{O}_3$, a CRF model is trained on the set consisting of the observed sequences $\mathcal{O}_1, \mathcal{O}_2, \mathcal{O}_4, ..., \mathcal{O}_M$, and the corresponding state sequences $\mathcal{S}_{1,\text{init}}, \mathcal{S}_{2,\text{init}}, \mathcal{S}_{4,\text{init}}, ..., \mathcal{S}_{M,\text{init}}$ obtained from the initial selection of the candidate key points. The set of parameters Λ_3, estimated during the training phase using (4.23) and (4.24), is afterward employed to infer the sequence of states $\mathcal{S}_3$ which maximizes the conditional probability $\mathcal{P}(\mathcal{S}_3 | \mathcal{O}_3, \Lambda_3)$. This procedure is repeated for each observed sequence from the dataset.

The inference problem of interest here consists of finding the most probable sequence of labels for an unlabeled observation sequence. The problem is typically solved either by the Viterbi method or by the method of maximal marginals (Rabiner, 1989). The maximal marginals approach is adopted in this work. This approach is further adapted for the task at hand by introducing additional constraints for the state transitions. Namely, the hidden state sequences in the presented CRF structure are defined such that they start with the state 1 and are ordered consecutively in a left-right structure, meaning that the state 1 can either continue with a self-transition or transition to the state 2. Furthermore, some states might not appear in all sequences, for example, a sequence can transition from the state 5 directly to the state 7, without occurrence of the state 6. By analogy with the left-right Bakis topology in HMMs, the contributions from the potential functions which correspond to the transitions to the past states (i.e., $\exp\left(\bar{\varphi}_{i,j}\Phi_{i,j}(s_k,s_{k-1},\mathcal{O},t_k)\right)$ in (4.20) for $i<j$ and $i,j\in\{1,N_s\}$) are minimized by setting low values for the parameters $\bar{\varphi}_{i,j}$. Thereby, the transitions to the previously visited states in the model are less likely to occur. Additionally, another constraint is introduced that minimizes the possibility of transitioning to distant future states, by setting low values for the parameters $\bar{\varphi}_{i,j}$ for $j>i+2$. As a result, the potential functions corresponding to the transitions to more than two future states have approximately zero values. These constraints reflect the sequential ordering of the hidden states in the CRF model for the considered problem.

The estimation of the hidden state at time t_k is based on computing the maximal marginal probabilities of the graph edges, and are solved by the forward–backward algorithm

$$P(s_k,s_{k-1}|\mathcal{O},t_k)\propto\bar{\alpha}_{k-1}(s_{k-1})\Omega(s_k,s_{k-1},\mathcal{O},t_k)\bar{\beta}_k(s_k),\qquad(5.18)$$

where $\bar{\alpha}$ and $\bar{\beta}$ denote the forward and backward variables which are calculated in the same recursive manner as with HMMs (Section 4.2.1). The functions Ω_k in (5.18) correspond to the transition feature functions in (4.20) (Sutton and McCallum, 2006), that is,

$$\Omega_k(s_k,s_{k-1},\mathcal{O},t_k)=\exp\left\{\sum_{l_1}\bar{\varphi}_{l_1}\Phi_{l_1}(s_k,s_{k-1},\mathcal{O},t_k)\right\}\qquad(5.19)$$

for $l_1\in\left\{1,N_s^2\right\}$.

The classification rate comparing the initial sequence of states $\mathcal{S}_{m,\text{init}}$ and the sequence of states obtained by the CRF labeling $\mathcal{S}_m$ is used as a measure of fit for each trajectory m

$$\Xi=\frac{1}{T_m}\sum_{k=1}^{T_m}I\left(s_{m,k,\text{init}}=s_{m,k}\right).\qquad(5.20)$$

The generation of a trajectory for reproduction of the demonstrations is analogous to the generalization procedure for HMM described in Section 5.2.2.3. To recapitulate, for each sequence of hidden states, the key points are taken to be the transitions between the states. The temporal variance across the set of sequences is tackled by employing the multidimensional DTW algorithm. The trajectory X_φ corresponding to the states sequence with the maximum classification rate $\mathcal{S}_\varphi$ in (5.20) is selected as a reference sequence, against which all the other trajectories are aligned. This process results in temporally warped key points sequences K_m^{DTW}, where the key points' time stamps correspond to the time vector of the reference trajectory X_φ. Weighting coefficients are introduced based on the variance of key points across the demonstrations associated with the same time stamps, in order to account for the relative importance for interpolation. The key points from all trajectories are afterward interpolated with a smoothing cubic spline, yielding a generalized trajectory X_{gen}.

5.2.3.4 Experiments

The presented concept for robot PbD is assessed through simulations of two types of manufacturing processes: painting and shot peening. Perception of the demonstrated trajectories is performed with an optical tracking device. The approach is implemented in MATLAB environment by using the CRF chain toolbox (Schmidt *et al.*, 2008).

5.2.3.4.1 Experiment 1

The experimental setup shown in Figure 5.18a is employed to simulate the trajectories of a tool for painting target objects. The painting task is demonstrated 14 times, and it consists of first painting the contour followed by painting the interior part of the panel. The demonstrations are shown in Figure 5.18c. The trajectories lengths T_m range from 1393 to 2073 time frames. For this task, the orientation of the tool is required to be normal to the panel during the painting; therefore, it is not considered as a relevant discriminative feature. For the initial selection of key points, the trajectories are automatically segmented with the LBG algorithm using 32 clusters of discrete vectors. A total of 33 transitions between the LBG's discrete symbols appeared in all trajectories, and these are adopted as the initial key points. The first and the last measurements of each trajectory are added as key points, resulting in 35 key points per trajectory. One sample trajectory with the identified key points is shown in Figure 5.18b. The demonstrated trajectories are discretized by utilizing the vectors consisting of normalized 3D positions and velocities

(a)

Figure 5.18 (a) Experimental setup for Experiment 1 showing the optical tracker, the tool with attached markers and the object for painting. (b) One of the demonstrated trajectories with the initially selected key points. The arrow indicates the direction of the tool's motion. (c) Demonstrated trajectories and the generalized trajectory.

(b)

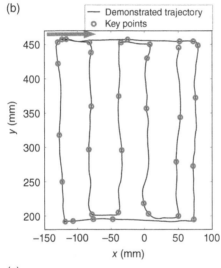

(c)

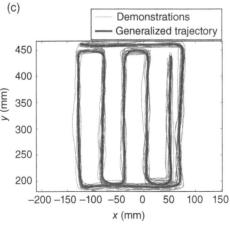

Table 5.4 Means and standard deviations of the classification rates obtained by CRF for the painting task in Experiment 1.

	Classification rates
1. Position	81.17 (±9.20)
2. Velocity	76.86 (±3.27)
3. Position and velocity	89.11 (±3.13)

$\boldsymbol{\alpha}_m(t_k) = \left[\hat{\mathbf{x}}_m(t_k)^T \ \ \hat{\mathbf{v}}_m(t_k)^T \right]^T$. With 32 symbols used per feature, the total number of observation symbols N_o amounted to 192.

The percentage of the average classification rates of CRFs from (5.20) calculated for different observation features are given in Table 5.4. The best classification rates are obtained when both velocities and positions are used as observation features. The generalized trajectory is shown by the thick line in Figure 5.18c, superimposed with the demonstrated trajectories. Note that even if some key points were wrongfully classified by the CRF, they would not have big impact on the generalization due to the introduced weighting scheme, which assigns low weights for fitting to the key points with high variability relative to the key points from the other trajectories.

5.2.3.4.2 Experiment 2

The second task for evaluation of the presented approach pertains to application of a shot peening procedure for surface finishing (Irish *et al.*, 2010). Shot peening is a process of impacting material's surface with small spherical elements called "shots." This process produces a compressive residual stress layer on material's surface, which increases the resistance to cracks caused by material fatigue or stress corrosion.

Demonstration of the peening task is performed seven times, with the number of measurements T_m varying between 2438 and 2655. Conversely to the painting process in Experiment 1, where the demonstrations were simulated in the university's lab, this task's demonstrations were captured for a process of peening in a real industrial environment. The trajectories represent waiving motions over a curved surface (Figure 5.19). Tool's positions, velocities, and orientations are examined as observation features. The number of clusters ρ for the initial key points selection with the LBG algorithm is set to 4. This choice is appropriate for the simple waving pattern of the task. A total of 41 key points per trajectory are identified, making the number of hidden states N_s equal to 40. One sample

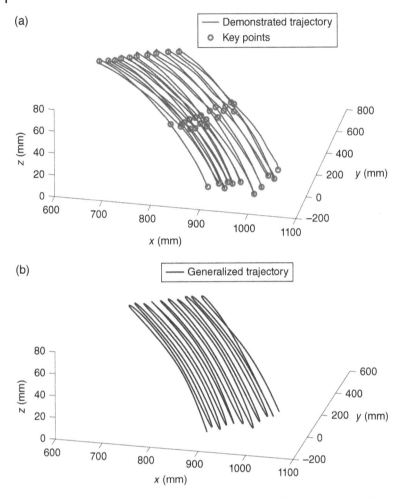

Figure 5.19 (a) Plot of a sample demonstrated trajectory for the peening task from Experiment 2 and a set of initially selected key points, and (b) generalized trajectory for the peening experiment.

trajectory with the initially selected key points is shown in Figure 5.19a. The discretization of the trajectories is performed with the number of pre-specified clusters for LBG equal to 16, that is, there are 16 discrete observation symbols for each dimension of the observed sequences.

The classification rates of the trajectories segmentation with CRF are given in Table 5.5. From the results provided, it can be concluded that the velocity features are the most suitable for CRF segmentation of this example. The reason behind it is that the key points corresponding to

Table 5.5 Means and standard deviations of the classification rates obtained by CRF for the peening task in Experiment 2.

	Classification rates
1. Position	79.06 (±4.33)
2. Velocity	94.30 (±1.79)
3. Position and velocity	94.70 (±1.55)
4. Position, velocity, and orientation	93.49 (±4.37)

the peaks and troughs of the traversed waving motions (Figure 5.19) are associated with the minimums of the trajectories velocities, whereas the middle parts of the waving motion corresponding to the waves' baseline are associated with the maximums velocities. Hence, these features are invariant for the entire set, and are informative for classification purposes. With regard to the tool's orientation, only the angles around the axis normal to the surface are taken into consideration, since the orientations around the other two axes are almost constant and do not convey important information for classification purposes. The results provided in Table 5.5 indicate that the orientation information does not improve the classification rates. Moreover, the CRF classification with the orientation as the only observation feature failed. The generalized trajectory for the case of position and velocity employed as observation features is shown in Figure 5.19b. The demonstrated trajectories are not shown in Figure 5.19b, since the demonstrations are too dissimilar, and it would be impossible to differentiate the individual demonstrations in a superimposed plot.

5.2.3.5 Comparisons with Related Work

The presented approach is compared to HMM, due to its wide usage for labeling and analysis of sequential data. To ensure that both methods are compared on an equal basis, a discrete form of HMM is employed, where the input data are the same discrete observation sequences $\mathcal{O}_1, \mathcal{O}_2, ..., \mathcal{O}_M$ and the label sequences $\mathcal{S}_{1,\text{init}}, \mathcal{S}_{2,\text{init}}, ..., \mathcal{S}_{M,\text{init}}$, which were used in the CRF approach. The multidimensional HMM reported by Yang *et al.* (1994) is implemented. Different from the 1D HMM, the multidimensional HMM has one observation matrix $\mathbf{B}^{(d)}$ for each dimension d of the observed data. The model is built on an assumption that each dimension of the observed sequences is independent from the other dimensions. On the other hand, the state transition matrix $\mathbf{A}$ has the same form as in the 1D case. The training and inference problems are solved with small modifications of the 1D HMM (see the work of Yang *et al.* (1994) for details). To maintain consistency with the adopted cross-validation testing for CRF, the HMM parameters are initialized using the CRF-labeled data for one of the trajectories (e.g., $\mathcal{S}_{3,\text{init}}$ and $\mathcal{O}_3$), followed by the model parameters estimation with the Baum–Welch algorithm for the rest of the trajectories (e.g., $\mathcal{O}_1,\mathcal{O}_2,\mathcal{O}_4...,\mathcal{O}_M$). The most probable sequence of hidden states for each observation trajectory and the given model are obtained by the Viterbi algorithm. This procedure is repeated for each of the observed trajectories.

For Experiment 1 related to the painting task, the mean classification rates for the multidimensional HMM obtained from (5.20) are reported in Table 5.6. For comparison, the classification rates obtained by the CRF approach from Table 5.4 are also provided in the rightmost column. For the three cases of observation features considered, the HMM produced lower number of correct labels than the CRF. The results from statistical analysis using paired *t*-tests showed statistically significant difference in the classification rates by the CRF and HMM at a 0.05 level. The following *p*-values were obtained for the different observation features: position ($p = 0.013$), velocity ($p = 0.026$), and position/velocity ($p < 0.01$).

For the peening task in Experiment 2, the classification rates obtained from both HMM and CRF are provided in Table 5.7. For all four types of observed features reported in the table, the HMM generates higher errors in predicting the labels. The differences in the classification performance of CRF and HMM methods are found to be statistically significant at a 0.05 level. The *p*-values of the performed paired *t*-tests are less than 0.01 for the different observed features in Table 5.7. It can be noted that adding the tools' orientation degrades significantly the HMM performance. When using both positions and velocities as observed features, the classification accuracy is improved, but is still lower when compared to the CRF success rates.

Table 5.6 Means and standard deviations of the classification rates obtained by HMM and CRF for the painting task from Experiment 1.

	Classification rates	
	HMM	CRF
1. Position	77.69 (±1.96)	81.17 (±9.2)
2. Velocity	74.75 (±1.99)	76.86 (±3.27)
3. Position and velocity	86.46 (±1.20)	89.11 (±3.13)

Table 5.7 Means and standard deviations of the classification rates obtained by HMM and CRF for the peening task from Experiment 2.

	Classification rates	
	HMM	CRF
1. Position	74.99 (±3.12)	79.06 (±4.33)
2. Velocity	80.97 (±1.79)	94.30 (±1.79)
3. Position and velocity	89.98 (±1.81)	94.70 (±1.55)
4. Position, velocity, and orientation	64.50 (±4.62)	93.49 (±4.37)

The results are consistent with the findings in other studies in the literature (Lafferty *et al.*, 2001; Vail *et al.*, 2007a; Martinez and Kragic, 2008), which report higher classification accuracy of CRF when compared to HMM.

5.3 Locally Weighted Regression

Local learning methods approximate the training data locally by using a set of functions, as opposed to the global regression methods which aim to calculate a single function to fit the training data at a global level. For instance, for a continuous input function y, a global regression approach calculates a continuous output function $\hat{y}$, to minimize a cost criterion,

$$C = \sum_i \mathcal{L}(f(\mathbf{x}_i, \theta), y_i), \tag{5.21}$$

where $\mathbf{x}_i$ denotes the ith vector of the input data, θ is a set of parameters of the model, $\hat{y} = f(\mathbf{x}_i, \theta)$ is the estimated function, and $\mathcal{L}(f(\mathbf{x}_i, \theta), y_i)$ is a loss function.

A common choice for the loss function is the least-squares difference, that is,

$$C = \sum_i (f(\mathbf{x}_i, \theta) - y_i)^2. \tag{5.22}$$

In local regression, the cost function is minimized locally, that is, $C(\mathbf{q})$ is a function of the input data around a query point $\mathbf{q}$. Examples of local learning models are nearest neighbor, weighted average, locally weighted regression (Atkeson *et al.*, 1997), and locally weighted projection regression (Vijayakumar and Schaal, 2000). Nearest-neighbor approach employs a distance function to find the nearest data within the training set to a query point, whereas weighted average, as the name implies, outputs the average of the data in the vicinity of a query point weighted by the distance between the data points and the query point. Locally weighted regression fits a surface through the data points in the vicinity of the query point by using a distance weighted regression. Locally weighted projection regression utilizes linear univariate local models by projecting the input data into a local plane.

Relevance of the data in locally learning models is defined using a distance function. A commonly used distance function is the Euclidean distance, where for a multidimensional data point $\mathbf{x} \in R^n$ and a query point $\mathbf{q} \in R^n$, the distance is defined as

$$d_E(\mathbf{x},\mathbf{q}) = \sqrt{(\mathbf{x}-\mathbf{q})^T(\mathbf{x}-\mathbf{q})} = \sqrt{\sum_j (x_j - q_j)^2}, \tag{5.23}$$

with j used to denote the index of the components of the vectors $\mathbf{x}$ and $\mathbf{q}$. Other distance functions that have found use in local function approximation include diagonally weighted Euclidean distance, where $\mathbf{M}$ is a diagonal matrix with the m_j elements acting as scaling factors for the jth dimension of the input data,

$$d_m(\mathbf{x},\mathbf{q}) = \sqrt{(\mathbf{x}-\mathbf{q})^T \mathbf{M}^T \mathbf{M}(\mathbf{x}-\mathbf{q})} = \sqrt{\sum_j m_j (x_j - q_j)^2}. \tag{5.24}$$

Also Mahalanobis distance is employed, which is similar to the previous distance function in (5.24); only the matrix $\mathbf{M}$ is allowed to have any arbitrary form, and it is not constrained to being a diagonal matrix,

$$d_M(\mathbf{x},\mathbf{q}) = \sqrt{(\mathbf{x}-\mathbf{q})^T \mathbf{M}^T \mathbf{M}(\mathbf{x}-\mathbf{q})} = \sqrt{\sum_j m_j (x_j - q_j)^2}. \tag{5.25}$$

Another commonly used function is Minkowski distance function, where, instead of using an L_2 norm, an arbitrary, weighted or unweighted, L_p norm of the distance to the query point is employed.

$$d_p(\mathbf{x},\mathbf{q}) = \left(\sum_j |x_j - q_j|^p \right)^{1/p}. \tag{5.26}$$

Weighting functions are employed to calculate the weight of the data point $\mathbf{x}$, based on a used distance function. These functions provide a smooth gradual decay with the increasing distance from the query point $\mathbf{q}$. In statistics, the weighting functions are called kernel functions, and correspondingly, the problem of fitting a locally weighted function is called kernel regression. The following several weighting functions have been employed:

1. Inverse function uses weighting based on the power of the distance to the query point, that is,

$$K(d(\mathbf{x},\mathbf{q})) = \frac{1}{d^p}. \tag{5.27}$$

2. Gaussian weighting function values the weights by calculating

$$K(d(\mathbf{x},\mathbf{q})) = e^{-d^2}. \tag{5.28}$$

3. Exponential weighting function,

$$K(d(\mathbf{x},\mathbf{q})) = e^{-|d|}. \tag{5.29}$$

4. Quadratic weighting function, given by

$$K(d(\mathbf{x},\mathbf{q})) = \begin{cases} (1-d^2) & \text{if } |d| < 1 \\ 0 & \text{otherwise} \end{cases}. \tag{5.30}$$

Similarly, a number of other weighting functions have been used; some of them developed as variants of the weighting functions given in (5.27)–(5.30).

Locally weighted regression employs weighting of the input data to calculate a continuous output function $\hat{y}$,

$$\hat{y}(\mathbf{q}) = \frac{\sum y_i K(d(\mathbf{x}_i,\mathbf{q}))}{\sum K(d(\mathbf{x}_i,\mathbf{q}))}, \tag{5.31}$$

where, as a reminder, $\mathbf{x}_i$ denotes the ith vector in the training input data.

Another approach for performing locally weighted regression is, instead of weighting the data points, to weight the training criterion, that is, the error in fitting the function, as in

$$C(\mathbf{q}) = \sum_i \left[(\hat{y}_i - y_i)^2 K(d(\mathbf{x}_i,\mathbf{q})) \right]. \tag{5.32}$$

In this case, the approximation function $\hat{y} = f(\mathbf{x}_i, \theta(\mathbf{q}))$ fits local models to the data, with the model parameters θ depending of the input data in the vicinity of the query point $\mathbf{q}$. Accordingly, the approximation function will produce different set of model parameters for different query points.

Linear models for locally weighted regression utilize a set of local models that are linear in the unknown parameters θ. The weighting functions in linear local models are commonly called basis functions. In the case of linear locally weighted regression, the cost function becomes

$$C(\mathbf{q}) = \sum_i \left[\left(x_i^T \theta - y_i \right)^2 K(d(\mathbf{x}_i, \mathbf{q})) \right]. \tag{5.33}$$

This approach allows fast approximation of nonlinear functions, where the computational cost increases linearly with the number of inputs. These advantages render it suitable for real-time applications.

The method of locally weighted regression has been applied in robotic learning from demonstration based on the dynamical systems approach described in Section 4.4. In this case, a set of Gaussian basis functions are employed to locally approximate a nonlinear force function F, related to a set of observed trajectories.

To learn a task from demonstrations, the nonlinear force function F is calculated from (4.26) by substituting the values for the position, velocity, and acceleration, denoted x_1, $\dot{x}_1$, and $\dot{x}_2$, of observed trajectories,

$$F_{\text{target}} = \tau \dot{x}_2 - \alpha \left(\beta \left(x_{\text{goal}} - x_1 \right) - \tau \dot{x}_1 \right). \tag{5.34}$$

Next, the phase variable φ is obtained by integrating the canonical system (4.27), and the weights w_i in (4.29) are calculated that minimize the cost function

$$C = \sum_{i=1}^{N} \left(F_{\text{target}} - F \right)^2, \tag{5.35}$$

by solving a linear regression problem. The weighting Gaussian functions are given in (4.30). Weighted Euclidean distance is employed as a distance function, where the standard deviation of the Gaussian basis functions is used as a weighting distance factor. Similarly, for learning motion primitives in rhythmic tasks, von Mises basis functions, given in (4.31), are adopted due to the periodic nature of the input trajectories.

To recreate a generalized trajectory that will converge toward a goal location x_{goal}, the weights w_i obtained by solving (5.35), and corresponding to the data for observed task trajectories, are reused. The value of the canonical variable φ is set to 1 at the beginning of the movement, and the canonical system (4.27) is integrated at each time step to obtain the current value of the variable, thus driving the value of the function F.

The method produces a smooth output trajectory that converges toward the discrete goal, with the task execution strategy being robust to perturbations to the state x.

5.4 Gaussian Process Regression

Within the published literature, a body of work employed Gaussian processes for robot leaning from human demonstrations (Shon *et al.*, 2005; Grimes *et al.*, 2006; Schneider and Ertel, 2010).

In general, Gaussian processes theory is used for learning the distribution of a regression function f given a set of input variables $x \in R^D$ and noisy observed target variables $y \in R$. The target variables are modeled as $y = f(x) + \varepsilon$, where ε is a random noise variable that is assumed to be independent and identically distributed with respect to the observed variables. The Gaussian process regression (GPR) model calculates a predictive distribution $y = \mathcal{P}(f^*|x^*, \mathcal{D})$, where x^* denotes a new input value, and $\mathcal{D}$ is a set of tuples consisting of n input points and related observations, $\mathcal{D} = \{(x_i, y_i)_{i=1}^n\}$. Gaussian processes consider the predicted values as samples from a zero-mean Gaussian distribution, resulting in a predictive Gaussian distribution $\mathcal{N}(\bar{f}^*, \text{cov}|f^*|)$ with mean (Schneider and Ertel, 2010)

$$\bar{f}^* = K^* \left(K + \sigma_n^2 I\right)^{-1} y, \tag{5.36}$$

and covariance matrix

$$\text{cov}|f^*| = K^{**} - K^* \left(K + \sigma_n^2 I\right)^{-1} K^{*T}. \tag{5.37}$$

In these equations σ_n^2 denotes the noise variance, whereas the covariance matrices K, K^*, K^{**} are defined by a covariance function, or kernel, $k(\cdot, \cdot)$, as $K_{i,j} = k(x_i, x_j)$, $K_{i,j}^* = k(x_i^*, x_j)$, and $K_{i,j}^{**} = k(x_i^*, x_j^*)$.

A variety of kernel functions are used in the literature, with the most common being the squared exponential kernel, also known as the radial basis function kernel, given with

$$k(x_p, x_q) = \sigma_f^2 \ \exp\left(-\frac{\|x_p - x_q\|^2}{2l^2}\right), \tag{5.38}$$

where σ_f^2 denotes the signal variance and l denotes the characteristic length scale. The set of parameters σ_n^2, σ_f^2, and l are referred to as hyper parameters of the covariance function, and their selection defines the form of the kernel. A standard approach for obtaining the values of the hyper parameters is by using the conjugate gradient algorithm to maximize the log likelihood of the data with respect to the hyper parameters. If the symbol ω substitutes the hyper parameters $\left(\sigma_n^2, \sigma_f^2, l\right)$, and $\mathbf{X} = \left\{(x_i)_{i=1}^n\right\}$ denotes the matrix of the input data, the log marginal likelihood is given by

$$\log \mathcal{P}(y|\mathbf{X}, \omega) = -\frac{1}{2}y^T K_y^{-1} y - \frac{1}{2}\log|K_y| - \frac{n}{2}\log 2\pi. \tag{5.39}$$

In (5.39), the notation $|\cdot|$ refers to determinant of a matrix, and $K_y = K + \sigma_n^2 I$.

A shortcoming of using GPR for real-world applications is the high computational cost of the algorithm. To reduce the computational expense, Schneider and Ertel (2010) proposed a local Gaussian process model in a PbD setting, where the prediction mean and the covariance are calculated as the Gaussian product of local model predictions,

$$z \cdot \mathcal{N}\left(\bar{f}^*, \text{cov}[f^*]\right) = \prod_{i=1}^L \mathcal{N}\left(\bar{f}_l^*, \sum_l = \text{diag}\left(\mathbb{V}[f^*]_l\right)\right), \tag{5.40}$$

where

$$\text{cov}[f^*] = \left(\sum_{l=1}^L \Sigma_l^{-1}\right)^{-1}, \bar{f}^* = \text{cov}[f^*]\left(\sum_{l=1}^L \Sigma_l^{-1}\bar{f}_l^*\right)^{-1}, \tag{5.41}$$

and z is the normalization constant. The approximate prediction function is then used for reconstructing an execution trajectory from a set of observed trajectories.

5.5 Summary

Approaches for generating a trajectory for task reproduction by a robot learner are reviewed in this chapter. GMR method employs a mixture of Gaussian density functions to calculate the conditional expectation of the temporal component of the demonstrated trajectories given the spatial component of the trajectory. Task planning with spline regression employs a set of trajectories key points, related to the transitions between the hidden states in the statistical models of HMM and CRF. Locally weighted regression is used for trajectory generation with tasks encoded via DMPs. GPR has also been employed for modeling a set of hidden and observed states from demonstration, followed by creating a task reproduction trajectory.

The chapter describes the entire procedure of transferring skills to robots for two industrial tasks: painting and shot peening. HMM and CRF are used for encoding of demonstrated tasks as a set of relevant key points, whereas DTW is employed for temporal clustering of the key points into a format that represents the generalized time flow of all demonstrations. A generalized trajectory for task reproduction is produced by interpolation of the key points across the demonstrated trajectories using spline regression.

References

Akaike, H., (1974). A new look at the statistical model identification. *IEEE Transactions on Automatic Control*, vol. **19**, no. 6, pp. 716–723.

Aleotti, J., and Caselli, S., (2006). Robust trajectory learning and approximation for robot programming by demonstration. *Robotics and Autonomous Systems*, vol. **54**, no. 5, pp. 409–413.

Asfour, T., Gyarfas, F., Azad, P., and Dillmann, R., (2006). Imitation learning of dual-arm manipulation tasks in a humanoid robot. *Proceedings of the International Conference on Humanoid Robots*, Genoa, Italy, pp. 40–47.

Atkeson, C.G., Moore, A.W., and Schaal, S., (1997). Locally weighted learning for control. *Artificial Intelligence Review*, vol. **11**, no. 1–5, pp. 75–113.

Berkhin, P., (2001). Survey of clustering data mining techniques. *Grouping Multidimensional Data*, Springer, pp. 25–71.

Bezdek, D., (1981). *Pattern Recognition with Fuzzy Objective Function Algorithms*. New York, USA: Plenum Press.

Billard, A., Eppars, Y., Calinon, S., Schaal, S., and Cheng, G., (2004). Discovering optimal imitation strategies. *Robotics and Autonomous Systems*, vol. **47**, no. 2–3, pp. 69–77.

Calinon, S., (2009). *Robot Programming by Demonstration: A Probabilistic Approach*. Boca Raton, USA: EPFL/CRC Press.

Calinon, S., and Billard, A., (2004). Stochastic gesture production and recognition model for a humanoid robot. *Proceedings of IEEE/RSJ International Conference on Intelligent Robots and Systems*, Sendai, Japan, pp. 2769–2774.

Calinon, S., and Billard, A., (2007). Learning of gestures by imitation in a humanoid robot. *Imitation and Social Learning in Robots, Humans and Animals: Social and Communicative Dimension*. (Eds.) Dautenhahn, K., and Nehaniv, C.L., Cambridge, USA: Cambridge University Press, pp. 153–178.

Calinon, S., D'halluin, F., Sauser, E.L., Caldwell, D.G., and Billard, A.G., (2010). Learning and reproduction of gestures by imitation: an approach based on hidden Markov model and Gaussian mixture regression. *IEEE Robotics and Automation Magazine*, vol. **17**, no. 2, pp. 44–54.

Davies, D.L., and Bouldin, D.W., (1979). A cluster separation measure. *IEEE Transactions on Pattern Analysis and Machine Intelligence*, vol. **1**, no. 2, 224–227.

DeCaprio, D., Vinson, J.P., Pearson, M.D., Montgomery, P., Doherty, M., and Galagan, J.E., (2007). Conrad: Gene prediction using conditional random fields, *Genome Research*, vol. **17**, no. 9, pp. 1389–1396.

Gribovskaya, E., and Billard, A., (2008). Combining dynamical systems control and programming by demonstration for teaching discrete bimanual coordination tasks to a humanoid robot. *Proceedings of ACME/IEEE International Conference on Human-Robot Interaction*, Amsterdam, the Netherlands, pp. 1–8.

Grimes, D.B., Chalodhorn, R., and Rao, R.P.N., (2006). Dynamic imitation in a humanoid robot through nonparameteric probabilistic inference. *Proceedings of Robotics: Science and Systems*, Cambridge, USA: MIT Press.

He, X., Zemel, R.S., and Carreira-Perpinan, M.A., (2004). Multiscale conditional random fields for image labeling. *Proceedings of IEEE Computer Society Conference on Computer Vision and Pattern Recognition*, Washington, USA, pp. 695–702.

Inamura, T., Tanie, H., and Nakamura, Y., (2003). Keyframe compression and decompression for time series data based on the continuous hidden Markov model. *Proceedings of IEEE/RSJ International Conference on Intelligent Robots and Systems*, Las Vegas, USA, pp. 1487–1492.

Inamura, T., Kojo, N., and Inaba, M., (2006). Situation recognition and behavior induction based on geometric symbol representation of multimodal sensorimotor patterns. *Proceedings of IEEE/RSJ International Conference on Intelligent Robots and Systems*, Beijing, China, pp. 5147–5152.

Irish, A., Mantegh, I., and Janabi-Sharifi, F., (2010). A PbD approach for learning pseudo-periodic robot trajectories over curved surfaces. *Proceedings of IEEE/ASME International Conference on Advanced Intelligent Mechatronics*, Montréal, Canada, pp. 1425–1432.

Kass, R.E., and Raftery, A.E., (1995). Bayes factors. *Journal of the American Statistical Association*, vol. **90**, no. 430, pp. 773–795.

Kaufman, L., and Rousseeuw, P., (1990). *Finding Groups in Data: An Introduction to Cluster Analysis*. New York, USA: John Wiley & Sons, Inc.

Kjellstrom, H., Romero, J., Martinez, D., and Kragic, D., (2008). Simultaneous visual recognition of manipulation actions and manipulated objects. *Proceedings of European Conference on Computer Vision, Part II*, LNCS 5303, pp. 336–349.

Klinger, R., Furlong, L.I., Friedrich, C.M., Mevissen, H.T., Fluck, J., Sanz, F., *et al.*, (2007). Identifying gene specific variations in biomedical text. *Journal of Bioinformatics and Computational Biology*, vol. **5**, no. 6, pp. 1277–1296.

Lafferty, J., McCallum, A., and Pereira, F., (2001). Conditional random fields: probabilistic models for segmenting and labeling sequence data. *Proceedings of International Conference on Machine Learning*, Williamstown, USA, pp. 282–289.

Lee, H.-K., and Kim, J.H., (1999). An HMM-based threshold model approach for gesture recognition. *IEEE Transactions on Pattern Analysis and Machine Intelligence*, vol. **21**, no. 10, pp. 961–973.

Lee, D., and Nakamura, Y., (2006). Stochastic model of imitating a new observed motion based on the acquired motion primitives. *Proceedings of IEEE/RSJ International Conference on Intelligent Robots and Systems*, Beijing, China, pp. 4994–5000.

Linde, Y., Buzo, A., and Gray, R.M., (1980). An algorithm for vector quantizer design. *IEEE Transactions on Communications*, vol. **28**, no. 1, pp. 84–95.

Liu, G., Zhang, J., Wang, W., and McMilan, L., (2005). A system for analyzing and indexing human-motion databases. *Proceedings of International Conference on Management of Data*, Baltimore, USA, pp. 924–926.

Martinez, D., and Kragic, D., (2008). Modeling and recognition of actions through motor primitives. *Proceedings of IEEE International Conference on Robotics and Automation*, Pasadena, USA, pp. 1704–1709.

McDonald, R., and Pereira, F., (2005). Identifying gene and protein mentions in text using conditional random fields. *BMC Bioinformatics*, vol. **6**, no. S6, pp. 1–3.

Murphy, K., (1998). *Hidden Markov Model (HMM) Toolbox for MATLAB*. Available from: http://www.cs.ubc.ca/~murphyk/Software/HMM/hmm. html (accessed on September 14, 2016).

Pook, P.K., and Ballard, D.H., (1993). Recognizing teleoperated manipulations. *Proceedings of IEEE International Conference on Robotics and Automation*, Atlanta, USA, pp. 578–585.

Quattoni, A., Collins, M., and Darrell, T., (2005). Conditional random fields for object recognition. *Advances in Neural Information Processing Systems*, Cambridge, USA: MIT Press, pp. 1097–1104.

Rabiner, L., (1989). A tutorial on hidden Markov models and selected applications in speech recognition. *Proceedings of the IEEE*, vol. **77**, no. 2, pp. 257–286.

Rice, J., and Rosenblatt, M., (1983). Smoothing splines: regression, derivatives and deconvolution. *The Annals of Statistics*, vol. **11**, no. 1, pp. 141–156.

Rissanen, J., (1978). Modeling by shortest data description. *Automatica*, vol. **14**, no. 5, pp. 465–658.

Ross, D.A., Osindero, S., and Zemel, R.S., (2006). Combining discriminative features to infer complex trajectories. *Proceedings of International Conference on Machine Learning*, Pittsburgh, USA, pp. 761–768.

Schmidt, M., Swersky, K., and Murphy, K., (2008). *Conditional Random Field Toolbox for MATLAB*. Available from: http://www.cs.ubc.ca/~murphyk/Software/CRF/crf.html (accessed on September 14, 2016)

Schneider, M., and Ertel, W., (2010). Robot learning by demonstration with local Gaussian process regression. *Proceedings of IEEE/RAS International Conference on Intelligent Robots and Systems*, Taipei, Taiwan, pp. 255–260.

Schwarz, G., (1978). Estimating the dimension of a model. *Annals of Statistics*, vol. **6**, no. 2, pp. 461–464.

Sha, F., and Pereira, F., (2003). Shallow parsing with conditional random fields. *Proceedings of Conference of the North American Chapter of the Association for Computational Linguistics on Human Language Technology*, Stroudsburg, USA, pp. 213–220.

Shon, A.P., Grochow, K., and Rao, R.P.N., (2005). Robotic imitation from human motion capture using Gaussian processes. *Proceedings of Fifth IEEE-RAS International Conference on Humanoid Robots*, Tsukuba, Japan, pp. 129–134.

Sminchisescu, C., Kanaujia, A., Li, Z., and Metaxas, D., (2005). Conditional models for contextual human motion recognition. *Proceedings of IEEE International Conference on Computer Vision*, Beijing, China, pp. 1808–1815.

Smyth, P., (1998). Model selection for probabilistic clustering using cross-validated likelihood, *UCI-ICS Technical Report 98-09*, University of California, Irvine, USA.

Sutton, C., and McCallum, A., (2006). *An Introduction to Conditional Random Fields for Relational Learning*. (Eds.) Getoor, L., and Taskar, B., Cambridge, USA: MIT Press.

Tso, S.K., and Liu, K.P., (1997). Demonstrated trajectory selection by hidden Markov model. *Proceedings of International Conference on Robotics and Automation*, Albuquerque, USA, pp. 2713–2718.

Vail, D.L., Lafferty, J.D., and Veloso, M.M., (2007a). Feature selection in conditional random fields for activity recognition. *Proceedings of IEEE/RSJ International Conference on Intelligent Robots and Systems*, San Diego, USA, pp. 3379–3384.

Vail, D.L., Veloso, M.M., and Lafferty, J.D., (2007b). Conditional random fields for activity recognition. *Proceedings of International Conference on Autonomous Agents and Multi-agent Systems*, Honolulu, USA, pp. 235–243.

Vakanski, A., Janabi-Sharifi, F., Mantegh, I., and Irish, A., (2010). Trajectory learning based on conditional random fields for robot programming by demonstration. *Proceedings of IASTED International Conference on Robotics and Applications*, Cambridge, USA, pp. 401–408.

Vakanski, A., Mantegh, I., Irish, A., and Janabi-Sharifi, F., (2012). Trajectory learning for robot programming by demonstration using hidden Markov model and dynamic time warping. *IEEE Transactions on Systems, Man, and Cybernetics—Part B: Cybernetics*, vol. **44**, no. 4, pp. 1039–1052.

Vijayakumar, S., and Schaal, S., (2000). Locally weighted projection regression: an $O(n)$ algorithm for incremental real time learning in high dimensional space. *Proceedings of International Conference on Machine Learning*, San Francisco, USA, pp. 1079–1086.

Yang, J., Xu, Y., and Chen, C.S., (1994). Hidden Markov model approach to skill learning and its application to telerobotics. *IEEE Transactions on Robotics and Automation*, vol. **10**, no. 5, pp. 621–631.

Yang, J., Xu, Y., and Chen, C.S., (1997). Human action learning via hidden Markov model. *IEEE Transactions on Systems, Man, and Cybernetics—Part A*, vol. **27**, no. 1, pp. 34–44.

Zhou, F., De la Torre, F., and Hodgins, J.K., (2008). Aligned cluster analysis for temporal segmentation of human motion. *Proceedings of IEEE Conference on Automatic Face and Gestures Recognition*, Amsterdam, the Netherlands, pp. 1–7.

6

Task Execution

This chapter focuses on endowing robustness in the task execution step using vision-based robot control. A framework that performs all the steps of the learning process in the image space of a vision sensor is presented. The steps of task perception, task planning, and task execution are formulated with the goal of improved overall system performance and enhanced robustness to modeling and measurement errors. The constraints related to vision-based controller are incorporated into the trajectory learning process, to guarantee feasibility of the generated plan for task execution. Hence, the task planning is solved as a constrained optimization problem, with an objective to optimize a set of trajectories of scene features in the image space of a vision camera.

6.1   Background and Related Work

The developed approach for robot programming by demonstration (PbD) is based on the following postulations: (i) perception of the demonstrations is performed with vision cameras; and (ii) execution of the learned strategies is conducted using visual feedback from the scene.

The most often used perception sensors in the PbD systems are the electromagnetic (Dillmann, 2004), inertial (Calinon, 2009) and optical marker-based sensors (Vakanski *et al.*, 2010). However, attaching sensors on workpieces, tools, or other objects for demonstration purposes is impractical, tiresome, and for some tasks, impossible. This work concentrates on using vision sensors for perception of demonstrated actions in a PbD framework, due to the nonintrusive character of the measurements. Also, visual perception is the most acceptable sensing modality for the next generation of intelligent robots, through the usage of "robots' eyes"

Robot Learning by Visual Observation, First Edition. Aleksandar Vakanski and Farrokh Janabi-Sharifi.
© 2017 John Wiley & Sons, Inc. Published 2017 by John Wiley & Sons, Inc.

(probably in combination with the information gathered by other cameras or sensors located on robot's structure).

Another important aspect of the PbD process that has received little attention is the step of task execution, being often considered as an independent step from the task planning, for which it is assumed that different types of robot control techniques can be employed. In practice, however, the type of controller used for reproduction of a planned strategy can impose certain constraints on the system, which might render the robot performance unsuitable for achieving the task goals. For example, the required command velocities and/or accelerations might not be realizable by the robot learner, or the required motions might violate the robot's dexterous workspace limits. Therefore, in this work, both the task planning and the task execution (control) are considered synergistically.

From the control perspective, the uncertainties in the execution step can cause incorrect positioning of the robot's end point with respect to the scene objects. For instance, joints' gear backlashes or slippages, bending of the robot links, poor fixturing of the objects, or incorrect estimation of the poses of scene objects from noisy sensor measurements, can all have unfavorable effects on system's performance. Under these circumstances, the execution of the generated PbD strategy from the planning phase will rely on open-loop kinematic chain to position the end-effector, and hence it can fail to correctly reproduce the desired robot configurations. To address the robustness of robots' positioning under uncertainties during the task execution, a vision-based control strategy (i.e., visual servoing) is employed here (Hutchinson *et al.*, 1996; Janabi-Sharifi, 2002; Chaumette and Hutchinson, 2006). At present, the use of visual feedback information in the PbD literature is mainly limited to look-and-move control, meaning that vision cameras are only used for extracting the positions and/or orientations of objects in the scene, without being used directly for real-time control of robots' motion.

The approach presented employs a set of demonstrations captured as trajectories of relevant scene features projected onto the image plane of a stationary camera. A Kalman smoother (Rauch *et al.*, 1965) is initially employed to extract a generalized trajectory for each image feature. This set of trajectories represents smooth and continuous averages of the observed feature trajectories, and it is going to be used as reference trajectories in generating a plan for task reproduction. Similarly, a Kalman smoother is employed for obtaining reference velocities of the tracked object from the demonstrations. The planning step is formulated as an optimization problem, with a cost function that minimizes the distances between the current and reference image feature vectors and the current and reference object velocities. The constraints in the optimization model include visual, workspace, and robot constraints. All the constraints are formulated in a linear or conic form, thus enabling to solve the model as a convex optimization problem. The main advantage of employing convex optimization is the global convergence within the set of feasible solutions. Subsequently, an image-based visual servoing (IBVS) controller is employed to ensure robust execution of the generated feature trajectories in presence of uncertainties, such as image noise and camera modeling errors (Nematollahi *et al.*, 2012). Note that instead of Kalman smoothing, the initial generalization can be performed by other learning methods, such as the ones using hidden Markov model (HMM) or conditional random field (CRF) described in Chapter 4, or alternatively it can be performed by employing standard curve smoothing techniques. The Kalman smoothing algorithm was selected in this work as a trade-off between the computational speed of the smoothing algorithms and the generalization abilities of the learning algorithms.

The motivations for implementing image-based learning approach are multifold, as follows:

- Integration of a vision-based control strategy into a PbD framework for robust task execution
- Formulation of the planning process as a convex optimization problem to incorporate important task constraints, which are often neglected in the PbD domain
- Introduction of a unique framework for PbD learning with all the steps of the PbD process, that is, observation, planning, and execution steps, taking place in the image space (of a robot's vision system)

Only a few works in the literature deal with the problem of image-based control in the robot learning context. Asada *et al.* (2000) addressed the problem of different viewpoints of demonstrated actions between a teacher and a learner agent. In this work, two pairs of stereo cameras were used separately for observation of the demonstrations and for reproduction of the planned strategies, and a PA-10 Mitsubishi robot was employed both as a demonstrator and a learner. This approach avoids the three-dimensional (3D) reconstruction of the observed motions, by proposing a direct reconstruction of the demonstrated trajectory in the image plane of the learner. The epipolar constraint of stereo cameras was used for this purpose, which dictates that the projections of a point onto stereo cameras lay on the epipolar lines. Visual servoing control was employed afterward for execution of the projected trajectory onto the image plane of the learner entity. Different from the approach presented in this chapter, the work by Asada *et al.* (2000) focuses on imitation of a single demonstrated trajectory, by tackling solely the problem of transforming a demonstrated trajectory from the demonstrated image space into the image space of the robot learner.

Jiang and Unbehauen (2002) and Jiang *et al.* (2007) presented iterative image-based learning schemes, where a single static camera was used for observation of a demonstrated trajectory. The later work employed neural networks to learn an approximation of the image Jacobian matrix along the demonstrated trajectories, whereas the former work approximated directly the control signal from the demonstrated data. The learning laws were designed to reduce iteratively the tracking errors in several repetitive reproduction attempts by a robot learner, leading to convergence of the reproduced trajectory toward the demonstrated trajectory. Visual servoing control was employed for execution of the task in the image plane of the camera. Similar to the previously mentioned work in Asada *et al.* (2000), the proposed schemes consider imitation of a single demonstrated trajectory. Other shortcomings encompass the lack of integration of task constraints, and the possibilities of features leaving the field of view (FoV) of the camera, especially during the first or second repetitive trials when the tracking errors can be significant.

A similar work that also implements learning of the image Jacobian matrix was proposed by Shademan *et al.* (2009). Image-based control was employed for execution of several primitive skills, which entailed pointing and reaching tasks. Numerical estimation of the Jacobian matrix was established via locally least-square estimation scheme. However, this approach is limited to the learning of only several primitive motions, without providing directions of how it can be implemented for learning complex trajectories from demonstrations.

On the other hand, a large body of work in the literature is devoted to the path planning in visual servoing (Mezouar and Chaumette, 2002; Deng *et al.*, 2005; Chesi and Hung, 2007). The majority of these methods utilize an initial and a desired image of the scene, and devise a plan for the paths of salient image features in order to satisfy certain task constraints. Different from these works, the planning step in the framework presented here is initialized by multiple examples of the entire image feature trajectories that are acquired from the demonstrations. Additionally, the planning in this work is carried out directly in the image plane of a vision sensor. However, since direct planning in the image space can cause suboptimal trajectories of the robot's end point in the Cartesian space, a constraint is formulated in the model that forces the respective Cartesian trajectory to stay within the envelope of the demonstrated motions.

6.2 Kinematic Robot Control

Robot control involves establishing the relationship between the joint angles and the position and orientation (pose) of the robot's end point. From this aspect, for a given vector of robot joint angles $\mathbf{q} \in \mathcal{R}^{N_q}$, the forward robot kinematics calculates the pose of the end-effector (Sciavicco and Siciliano, 2005). Denavit–Hartenberg convention is a common technique for assigning coordinate frames to a robot's links, as well as for calculating the forward kinematics as a product of homogenous transformation matrices related to the coordinate transformation between the robot links frames

$$\mathbf{T}_e^b(\mathbf{q}) = \begin{bmatrix} \mathbf{R}_e^b(\mathbf{q}) & \mathbf{P}_e^b(\mathbf{q}) \\ \mathbf{0}_{1 \times 3} & 1 \end{bmatrix}. \tag{6.1}$$

In the given equation, $\mathbf{R}_e^b(\mathbf{q})$ and $\mathbf{P}_e^b(\mathbf{q})$ denote the rotation matrix and the position vector, respectively, of the end-effector frame with respect to the robot base frame.

The reverse problem of finding the set of joint angles corresponding to a given pose of the end-effector is called robot inverse kinematics. The equations of the inverse kinematics for a general robot type are nonlinear, and a closed-form solution does not exist. Robot control is therefore commonly solved using robot differential kinematics equations. In particular, the relation between the joint velocities and the resulting velocities of the end-effector is called forward differential kinematics, expressed by the following equation:

$$\mathbf{v}_e^b = \mathbf{J}(\mathbf{q}) \, \dot{\mathbf{q}}. \tag{6.2}$$

The matrix $\mathbf{J}(\mathbf{q})$ in (6.2) is called the robot Jacobian matrix, and provides a linear mapping between the joint and end-effector velocities. The forward differential kinematics can be easily derived for a generic robot type. The derivation of the equations for calculating the Jacobian matrix can be found in any textbook that covers robot modeling (e.g., Sciavicco and Siciliano, 2005).

The inverse differential kinematics problem solves the unknown set of joint velocities corresponding to given end-effector velocities. It is obtained by calculating the inverse of the Jacobian matrix in (6.2), that is,

$$\dot{\mathbf{q}} = \mathbf{J}^{-1}(\mathbf{q}) \mathbf{v}_e^b, \tag{6.3}$$

under the assumption that the Jacobian matrix is square and nonsingular. Obtaining the joint angles for robot control purposes is typically based on numerical integration of (6.3) over time.

Note that the Jacobian matrix is of size $N_q \times 6$; therefore, the inverse of the Jacobian matrix will not exist for robots with number of joints different from 6. If $N_q > 6$, the inverse kinematics problem can be solved by using the right pseudoinverse of the Jacobian matrix, $\mathbf{J}^\dagger = \mathbf{J}^T \left(\mathbf{J} \mathbf{J}^T \right)^{-1}$. In this case, the solution to (6.2) is

$$\dot{\mathbf{q}} = \mathbf{J}^\dagger(\mathbf{q})\mathbf{v}_e^b + \left(\mathbf{I} - \mathbf{J}(\mathbf{q})^\dagger\mathbf{J}(\mathbf{q})\right)\mathbf{b}, \tag{6.4}$$

where $\mathbf{b} \in \mathcal{R}^{N_q}$ is any arbitrary vector.

The joint configurations for which the Jacobian matrix is singular are called robot kinematic singularities, and the corresponding configurations will be singular configurations. These configurations may lead to unstable behavior, for example, unbounded joint velocities, and should be avoided if possible.

6.3 Vision-Based Trajectory Tracking Control

Visual servoing systems are generally categorized into three main servoing structures (Chaumette and Hutchinson, 2006): position-based visual servoing (PBVS), image-based visual servoing (IBVS), and hybrid visual servoing (HVS). With regards to the position of the camera(s) of the visual servoing systems, there are two configurations: camera mounted on robot's end-effector (a.k.a. camera-in-hand or eye-in-hand), and camera in a fixed position with respect to the end-effector (camera-off-hand).

6.3.1 Image-Based Visual Servoing (IBVS)

IBVS is based on positioning of the end-effector with respect to target objects by using feature parameters in the image space. Coordinates of several points from the target object can be considered as image feature parameters. Therefore, let the coordinates of n object feature points in the sworld frame are denoted with (X_i, Y_i, Z_i), for $i = 1, 2, ..., n$. The perspective projections of the feature points in the image plane are $\mathbf{p}_i = [x_i \ y_i]^T \in \mathbb{R}^2$, where

$$\begin{aligned} x_i &= \frac{X_i}{Z_i} = \frac{(u - u_0)}{f\alpha} \\ y_i &= \frac{Y_i}{Z_i} = \frac{(v - v_0)}{f}. \end{aligned} \tag{6.5}$$

Here, f denotes the focal length of the camera, u and v are pixel coordinates of the corresponding image features, u_0 and v_0 are pixel coordinates of the principal point, and α denotes the aspect ratio of horizontal and vertical pixel dimensions. The set of all vectors of image feature parameters $\mathbf{s} = \begin{bmatrix} \mathbf{p}_1^T & \mathbf{p}_2^T & ... & \mathbf{p}_n^T \end{bmatrix}^T$ defines the image space $S \subseteq \mathbb{R}^{2n}$. Let $\mathbf{s}^*$ denote the vector of desired image feature parameters, which is constant for the task of positioning of the robot's end-effector with respect to a stationary target object.

The goal of IBVS is to generate a control signal, such that error in the image space $\mathbf{e}_i = \mathbf{s} - \mathbf{s}^*$ is minimized, subject to some constraints. For the case of an eye-in-hand configuration, it can be assumed that the camera frame coincides with the end-effector frame, and the relationship between the time change of camera velocity and the image feature parameters is given with

$$\dot{\mathbf{s}} = \mathbf{L}(\mathbf{s})\mathbf{v}_c, \tag{6.6}$$

where $\mathbf{v}_c$ denotes the spatial camera velocity in the camera frame, and the matrix $\mathbf{L}(\mathbf{s})$ is referred to as image Jacobian or interaction matrix. For a perspective projection model of the camera, the interaction matrix is given as (Chaumette and Hutchinson, 2006)

$$\mathbf{L}(\mathbf{s}) = \begin{bmatrix} -1/Z_1 & 0 & x_1/Z_1 & x_1 y_1 & -1-x_1^2 & y_1 \\ 0 & -1/Z_1 & y_1/Z_1 & 1+y_1^2 & -x_1 y_1 & -x_1 \\ \vdots & \vdots & \vdots & \vdots & \vdots & \vdots \\ -1/Z_n & 0 & x_n/Z_n & x_n y_n & -1-x_n^2 & y_n \\ 0 & -1/Z_n & y_n/Z_n & 1+y_n^2 & -x_n y_n & -x_n \end{bmatrix}. \tag{6.7}$$

Note that Z_i is the depth coordinate of the target feature i with respect to the camera frame, whose value has to be estimated or approximated for computation of the interaction matrix.

For a properly selected set of features, traditional control laws based on exponential decrease of the error in image space ensures local asymptotic stability of the system and demonstrates good performance. Compared to PBVS, IBVS is less sensitive to camera calibration errors, does not require pose estimation, and is not as computationally expensive. On the other hand, the control law with exponential error decrease forces the image feature trajectories to follow straight lines, which causes suboptimal Cartesian end-effector trajectories (Figure 6.1). That is especially obvious for tasks with rotational camera movements, when the camera retreats along the optical axis and afterward returns to the desired location (Chaumette, 1998).

6.3.2 Position-Based Visual Servoing (PBVS)

PBVS systems use information for feature parameters extracted from image space to estimate the pose of an end-effector with respect to a target object. The control is designed based on the relative error between

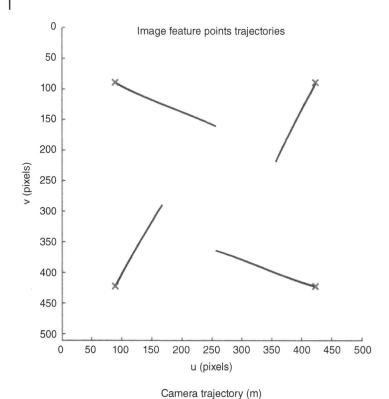

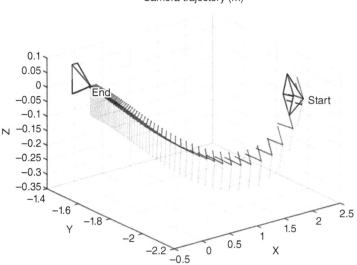

Figure 6.1 Response of classical IBVS: feature trajectories in the image plane, camera trajectory in Cartesian space, camera velocities, and feature errors in the image plane.

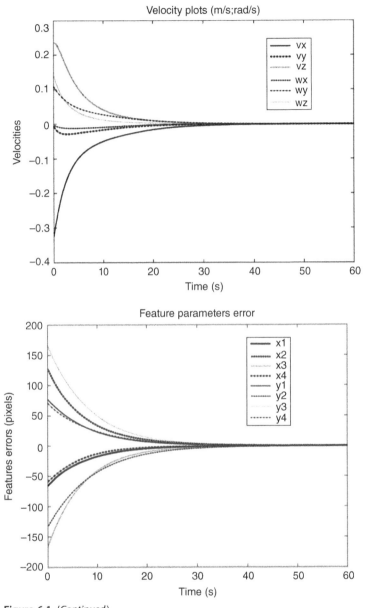

Figure 6.1 (*Continued*)

estimated and desired poses of the end-effector in the Cartesian space. There are several approaches for end-effector pose estimation in the literature, including close-range photogrammetric techniques (Yuan, 1989), analytical methods (Horaud *et al.*, 1989), least-squares methods (Liu *et al.*, 1990), and Kalman filter estimation (Wilson *et al.*, 2000; Ficocelli and Janabi-Sharifi, 2001).

Let $\mathbf{t}_c^{c^*}$ denotes the translation vector of the frame attached to the camera with respect to the desired camera frame, and let use angle/axis parameterization for the orientation of the camera with respect to the desired camera frame $\mathbf{R}_c^{c^*} = \theta\mathbf{u}$. The relative error between the pose of the current and desired pose of the camera is $\mathbf{e} = \left(\mathbf{P}_c^{c^*}, \theta\mathbf{u}\right)$. The relationship between the time change of the pose error and the camera velocity $\dot{\mathbf{e}} = \mathbf{L}_e\mathbf{v}_c$ is given by the interaction matrix

$$\mathbf{L}_e = \begin{bmatrix} \mathbf{R}_c^{c^*} & 0 \\ 0 & \mathbf{L}_{\theta\mathbf{u}} \end{bmatrix}, \tag{6.8}$$

where $\mathbf{I}_3$ is a 3×3 identity matrix, and matrix $\mathbf{L}_{\theta\mathbf{u}}$ is defined as (Malis *et al.*, 1999)

$$\mathbf{L}_{\theta\mathbf{u}} = \mathbf{I}_3 - \frac{\theta}{2}[\mathbf{u}]_{\times} + \left(1 - \frac{\mathrm{sinc}\theta}{\mathrm{sinc}^2(\theta/2)}\right)[\mathbf{u}]_{\times}^2, \tag{6.9}$$

where $\mathrm{sinc}\theta = (\sin\theta)/\theta$ and $\mathrm{sinc}0 = 1$. For the case of control with exponential decrease of the error (i.e., $\dot{\mathbf{e}} = -\lambda\mathbf{e}$), the camera velocity is

$$\mathbf{v}_c = -\lambda\mathbf{L}_e^{-1}\mathbf{e} = -\lambda \begin{bmatrix} \mathbf{R}_c^{c^*} & 0 \\ 0 & \mathbf{L}_{\theta\mathbf{u}}^{-1} \end{bmatrix} \begin{bmatrix} \mathbf{P}_c^{c^*} \\ \theta\mathbf{u} \end{bmatrix} = -\lambda \begin{bmatrix} \mathbf{R}_c^{c^*}\mathbf{P}_c^{c^*} \\ \theta\mathbf{u} \end{bmatrix}. \tag{6.10}$$

The final result in (6.10) follows from $\mathbf{L}_{\theta\mathbf{u}}^{-1}\theta\mathbf{u} = \theta\mathbf{u}$ (Malis *et al.*, 1999). The described control causes the translational movement of the camera to follow a straight line to the desired camera position, which is shown in Figure 6.2. On the other hand, the image features in Figure 6.2 followed suboptimal trajectories, and left the camera boundaries.

The advantages of PBVS stem from separation of the control design from the pose estimation, which endows the use of traditional robot control algorithms in the Cartesian space. This approach provides good control for large movements of the end-effector. However, the pose estimation is sensitive to camera and object model errors and it is computationally expensive. In addition, it does not provide a mechanism for regulation of features in the image space, which can cause features to leave the camera FoV (Figure 6.2).

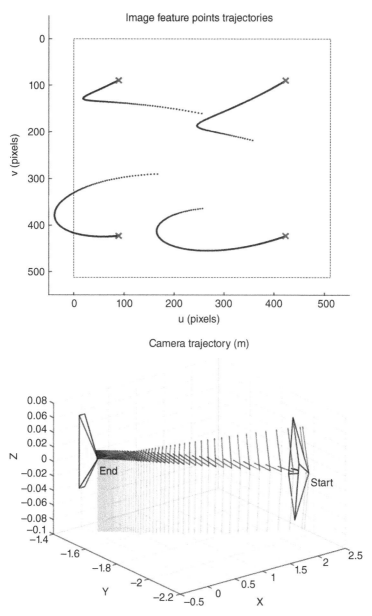

Figure 6.2 Response of classical PBVS: feature trajectories in the image plane, camera trajectory in Cartesian space, camera velocities, and feature errors in the image plane.

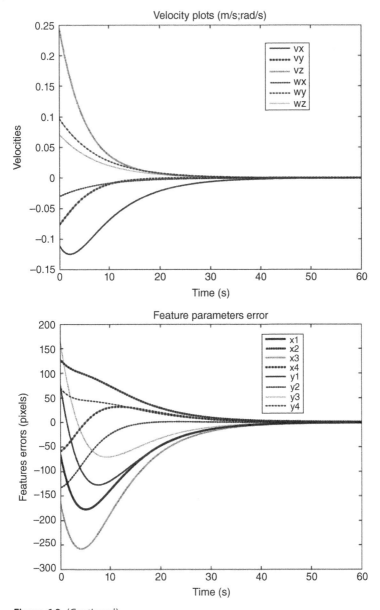

Figure 6.2 (*Continued*)

6.3.3 Advanced Visual Servoing Methods

Hybrid visual servo techniques have been proposed to alleviate the limitations of IBVS and PBVS systems. In particular, HVS methods offer a solution to the problem of camera retreat through decoupling of the rotational and translational degrees of freedom (DoFs) of the camera. The hybrid approach of 2-1/2-D (Malis *et al.*, 1999) utilizes extracted partial camera displacement from the Cartesian space to control the rotational motion of the camera, and visual features from the image space to control the translational camera motions. Other HVS approaches include the partitioned method (Corke and Hutchinson, 2001), which partitions the control of the camera along the optical axis from the control of the remained DoFs, and the switching methods (Deng *et al.*, 2005; Gans and Hutchinson, 2007), which use a switching algorithm between IBVS and PBVS depending on specified criteria for optimal performance. Hybrid strategies in both control and planning levels have also been proposed (Deng *et al.*, 2005) to avoid image singularities, local minima, and boundaries. Some of the drawbacks associated with most HVS approaches include the computational expense, possibility of features leaving the image boundaries, and sensitivity to image noise.

6.4 Image-Based Task Planning

6.4.1 Image-Based Learning Environment

The task perception step using vision cameras is treated in Section 2.2. The section stresses the importance for dimensionality reduction of the acquired image data through feature extraction.

Different camera configurations for observation of demonstrations have been used in the literature, for example, stereo pairs and multiple cameras. A single stationary camera positioned at a strategically chosen location in the scene is employed here. A graphical schematic of the environment is shown in Figure 6.3. It includes the stationary camera and a target object, which is manipulated either by a human operator during the demonstrations or by a robot during the task reproduction (or in the case discussed in Section 6.5.2, the object can be grasped by the robot's gripper but manipulated by a human teacher through kinesthetic demonstrations). The notation for the coordinate frames is introduced in Figure 6.3 as follows: camera frame $\mathcal{F}_c(O_c, x_c, y_c, z_c)$, object frame $\mathcal{F}_o(O_o, x_o, y_o, z_o)$, robot's end-point frame $\mathcal{F}_e(O_e, x_e, y_e, z_e)$, and robot base frame $\mathcal{F}_b(O_b, x_b, y_b, z_b)$. The robot's end-point frame is assigned to the central point at which the

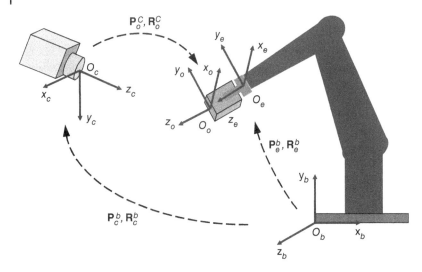

Figure 6.3 The learning cell, consisting of a robot, a camera, and an object manipulated by the robot. The assigned coordinate frames are: camera frame $\mathcal{F}_c(O_c, x_c, y_c, z_c)$, object frame $\mathcal{F}_o(O_o, x_o, y_o, z_o)$, robot base frame $\mathcal{F}_b(O_b, x_b, y_b, z_b)$, and robot's end-point frame $\mathcal{F}_e(O_e, x_e, y_e, z_e)$. The transformation between a frame i and a frame j is given by a position vector $\mathbf{P}_i^j$ and a rotation matrix $\mathbf{R}_i^j$.

gripper is attached to the flange of the robot wrist. The pose of a coordinate frame i with respect to a frame j is denoted with the pairs $\mathbf{P}_i^j, \mathbf{R}_i^j$ in the figure. As explained in Chapter 2, the image-based perception of demonstrations can provide estimation of the Cartesian pose of the object of interest by employing the homography matrix. Furthermore, the velocity of the object at given time instants is calculated by differentiating the pose.

6.4.2 Task Planning

The advantage of performing the perception, planning, and execution of skills directly in the image space of a vision camera is the endowed robustness to camera modeling errors, image noise, and robot modeling errors. However, planning in the image space imposes certain challenges, which are associated with the nonlinear mapping between the 2D image space and the 3D Cartesian space. More specifically, small displacements of the features in the image space can sometimes result in large displacements and high velocities of the feature points in the Cartesian space. In PbD context, this can lead to suboptimal Cartesian trajectories of the manipulated object. Hence, the manipulated object can leave the boundaries of the demonstrated space or potentially cause collisions with objects in the environment. Moreover, the velocities and accelerations required to achieve the planned motions of the manipulated object may be outside the limits of the motor abilities for the robot learner.

To tackle these challenges, the planning step in the image space is formalized here as a constrained optimization problem. The objective is to simultaneously minimize the displacements of features in the image space and the corresponding velocities of the object of interest in the Cartesian space, with respect to a set of reference image feature trajectories and reference object velocities. The optimization procedure is performed at each time instant of the task planning step. Euclidean norms of the changes in the image feature parameters and the Cartesian object velocities are employed as distance metrics, which are going to be minimized. Consequently, a second-order conic optimization model has been adopted for solving the problem at hand.

6.4.3 Second-Order Conic Optimization

Conic optimization is a subclass of convex optimization, where the objective is to minimize a convex function over the intersection of an affine subspace and a convex cone. For a given real-vector space Z, a convex real-valued function $\psi : \mathcal{K} \to \mathbb{R}$ defined on a convex cone $\mathcal{K} \subset Z$, and an affine subspace $\mathcal{H}$ defined by a set of affine constraints $h(\mathbf{z}) = 0$, a conic optimization problem consists in finding a point $\mathbf{z}$ in $\mathcal{K} \cap \mathcal{H}$, which minimizes the function $\psi(\mathbf{z})$. The second-order conic optimization represents a special case, when the convex cone $\mathcal{K}$ is a second-order cone.

A standard form of a second-order conic optimization problem is given as

$$\underset{\mathbf{z}}{\text{minimize }} \mathbf{c}^T \mathbf{z}$$

$$\text{subject to } \mathbf{Sz} = \mathbf{b} \tag{6.11}$$

$$\mathbf{z}_c \in \mathcal{K}$$

where the inputs are a matrix $\mathbf{S} \in \mathbb{R}^{D_1 \times D_2}$, vectors $\mathbf{c} \in \mathbb{R}^{D_2}$ and $\mathbf{b} \in \mathbb{R}^{D_1}$, and the output is the vector $\mathbf{z} \in \mathbb{R}^{D_2}$. The part of the vector $\mathbf{z}$ that corresponds to the conic constraints is denoted by $\mathbf{z}_c$, whereas the part that corresponds to the linear constraints is represented by $\mathbf{z}_l$, that is, $\mathbf{z} = \begin{bmatrix} \mathbf{z}_c^T & \mathbf{z}_l^T \end{bmatrix}^T$. For a vector variable $\mathbf{z}_{c,i} = \begin{bmatrix} z_{c,i}^1 & z_{c,i}^2 & \dots & z_{c,i}^{D_2} \end{bmatrix}$ that belongs to a second-order cone $\mathcal{K}$, one has $z_{c,i}^1 \leq \left\| \begin{bmatrix} z_{c,i}^2 & z_{c,i}^3 & \dots & z_{c,i}^{D_2} \end{bmatrix} \right\|$. As remarked before, $\|\cdot\|$ is used to denote Euclidean norm of a vector. The symbols D_1 and D_2 denote the dimensionality of the variables. An important property of the conic optimization problems is the convexity of the solutions space, that is, global convergence is warranted within the set of feasible solutions. To cast a particular problem into a second-order optimization form requires a mathematical model expressed through linear and second-order conic constraints.

6.4.4 Objective Function

The objective function of the optimization problem is formulated here as weighted minimization of distance metrics for the image feature trajectories and the object velocities with respect to a set of reference trajectories. The reference trajectories are generated by applying a Kalman smoothing algorithm, as explained in the following text.

For a set of M demonstrated trajectories represented by the image feature parameters $\left\{ \mathbf{r}_1^{(n)}, \mathbf{r}_2^{(n)}, \ldots, \mathbf{r}_M^{(n)} \right\}_{n=1}^{N}$, Kalman smoothers are used to obtain smooth and continuous average of the demonstrated trajectories for each feature point n. The observed state of each Kalman smoother is formed by concatenation of the measurements from all demonstrations. For instance, for the feature point 1, the observation vector at time t_k includes the image feature vectors from all M demonstrations, $\mathbf{o}_k^{(1)} = \left[\left(\mathbf{r}_1^{(1)}(t_k)\right)^T, \left(\mathbf{r}_2^{(1)}(t_k)\right)^T \ldots \left(\mathbf{r}_M^{(1)}(t_k)\right)^T \right]^T \in \mathbb{R}^{2M}$. The Kalman-smoothed trajectories for each feature point n are denoted by $\mathbf{r}^{(n), \text{ref}}$. Subsequently, the first part of the cost function is formulated to minimize the distance between an unknown vector related to the image feature parameters at the next time instant $\mathbf{r}^{(n)}(t_{k+1})$ and the reference image feature parameters at the next time instant, that is, $\left\| \mathbf{r}^{(n)}(t_{k+1}) - \mathbf{r}^{(n), \text{ref}}(t_{k+1}) \right\|$. The goal is to generate continuous feature trajectories in the image space, i.e., to prevent sudden changes in the image feature coordinates. To define the optimization over a conic set of variables, a set of auxiliary variables is introduced as

$$\tau_n \leq \left\| \mathbf{r}^{(n)}(t_{k+1}) - \mathbf{r}^{(n), \text{ref}}(t_{k+1}) \right\|, \quad \text{for } n = 1, 2, \ldots, N. \tag{6.12}$$

The first part of the objective function minimizes a weighted sum of the variables τ_n, based on the conic constraints (6.12).

The second part of the cost function pertains to the velocity of the target object. The objective is to ensure that the image feature trajectories are mapped to smooth and continuous velocities of the manipulated object. To retrieve the velocity of the object from camera-acquired images, the pose of the object $\left(\mathbf{P}_o^c, \mathbf{R}_o^c \right)$ is first extracted at each time instant by employing the homography transformation. By differentiating the pose, the linear and angular velocities of the object in the camera frame $\mathbf{v}_o^c(t_k) \in \mathbb{R}^3$, $\boldsymbol{\omega}_o^c(t_k) \in \mathbb{R}^3$ at each time instant for each demonstration are obtained (Sciavicco and Siciliano, 2005). Similar to the first part of the cost function, Kalman smoothers are used to generate smooth averages of the linear and angular velocities, that is, $v_o^{c,\,\mathrm{ref}} = \left(v_o^{c,\,\mathrm{ref}}, \omega_o^{c,\,\mathrm{ref}}\right)$. The objective is formulated to minimize the sum of Euclidean distances between an unknown vector related to the current linear and angular velocities, and the reference linear and angular velocities. By analogy to the first part of the cost function, two auxiliary conic variables are introduced

$$
\begin{aligned}
\tau_v &\le \left\| v_o^c(t_k) - v_o^{c,\,\mathrm{ref}}(t_k) \right\| \\
\tau_v &\le \left\| \omega_o^c(t_k) - \omega_o^{c,\,\mathrm{ref}}(t_k) \right\|
\end{aligned}
\tag{6.13}
$$

that correspond to the linear and angular object velocities, respectively.

The overall cost function is then defined as a weighted minimization of the sum of variables $\tau_1, ..., \tau_N, \tau_v, \tau_\omega$, that is,

$$
\mathrm{minimize} \left\{ \sum_{n=1}^{N} \alpha_n \tau_n + \alpha_v \tau_v + \alpha_\omega \tau_\omega \right\}
\tag{6.14}
$$

where the α's coefficients denote the weights of relative importance for the individual components in the cost function. The selection of the weighting coefficients, their influence on the performance of the system, and other implementation details, will be discussed in the later sections dedicated to validation of the method.

To recapitulate, when one performs trajectory planning directly in the image space for each individual feature point, the generated set of image trajectories of the object's feature points might not translate into a feasible Cartesian trajectory of the object. In the considered case, the reference image feature vectors obtained with the Kalman smoothing do not necessarily map into a meaningful Cartesian pose of the object. Therefore, the optimization procedure is performed to ensure that the model variables are constrained such that at each time instant there exists a meaningful mapping between the feature parameters in the image space and the object's pose in the Cartesian space.

Thus, starting from a set of reference feature parameters $\boldsymbol{\xi}^{\mathrm{ref}}(t_{k+1}) = \left\{ \mathbf{r}^{(n),\,\mathrm{ref}}(t_{k+1}) \right\}_{n=1}^{N}$ and reference velocity $\mathbf{v}_o^{c,\,\mathrm{ref}}(t_{k+1})$, the optimization will be performed at each time instant t_k to obtain a new optimal set of image feature parameters $\boldsymbol{\xi}(t_{k+1})$ which is close to the reference image feature parameters $\boldsymbol{\xi}^{\mathrm{ref}}(t_{k+1})$, and which entails feasible and smooth Cartesian object velocity $\mathbf{v}_o^c(t_k)$. From the robot control perspective, the goal is to find an optimal velocity of the end point (and subsequently, the velocity of object that is grasped by robot's gripper) $\mathbf{v}_o(t_k)$, which when applied at the current time will result in an optimal location of the image features at the next time step $\boldsymbol{\xi}(t_{k+1})$.

6.4.5 Constraints

The task analysis phase in the presented approach involves extraction of the task constraints from the demonstrated examples, as well as formulation of the constraints associated with the limitations of the employed robot, controller, and sensory system. Therefore, this section formulates the constraints in the optimization model associated with the visual space, the Cartesian workspace, and the robot kinematics. Note that the term "constraint" is used from the perspective of solving an optimization problem, meaning that it defines relationships between variables.

6.4.5.1 Image-Space Constraints

Constraint 1. The relationship between image feature velocities and the Cartesian velocity of the object can be expressed by introducing a Jacobian matrix of first-order partial derivatives

$$\dot{\boldsymbol{\xi}}(t) = \mathbf{L}(t)\mathbf{v}_o^c(t). \tag{6.15}$$

Since the optimization procedure is performed at discrete time intervals, an approximation of the continuous time equation (6.15) can be obtained by using Euler's forward discretization scheme

$$\boldsymbol{\xi}(t_{k+1}) = \boldsymbol{\xi}(t_k) + \mathbf{L}(t_k)\mathbf{v}_o^c(t_k)\Delta t_k, \tag{6.16}$$

where Δt_k denotes the sampling period at time t_k, whereas the matrix $\mathbf{L}(t_k)$ in the literature of visual servoing is often called image Jacobian matrix or interaction matrix (Chaumette and Hutchinson, 2006).

Constraint 2. The second constraint ensures that the image feature parameters in the next time instant $\mathbf{r}^{(n)}(t_{k+1})$ are within the bounds of the envelope of the demonstrated motions. For that purpose, first at each time step, the principal directions of the demonstrated motions are found by utilizing the eigenvectors of the covariance matrix from the demonstrations. For instance, for feature 1, the covariance matrix at each time instant is associated with the concatenated observation vectors from the entire set of demonstrations $m = 1, 2, ..., M$, i.e., $\text{cov}\left(\mathbf{r}_1^{(1)}(t_k), \mathbf{r}_2^{(1)}(t_k), ..., \mathbf{r}_M^{(1)}(t_k)\right)$. For the set of three trajectories in Figure 6.4a, the eigenvectors $\hat{\mathbf{e}}_1, \hat{\mathbf{e}}_2$ are illustrated at three different time instants t_k for $k = 10, 30, 44$. The observed image plane features are depicted by different types of markers in Figure 6.4.

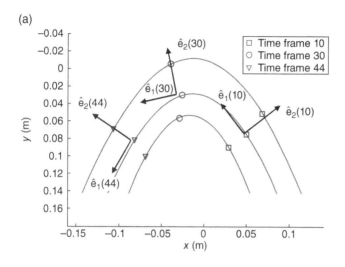

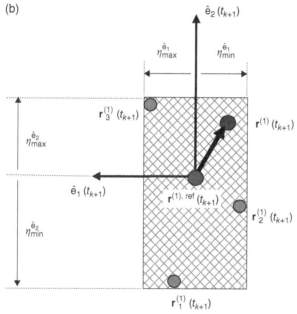

Figure 6.4 (a) The eigenvectors of the covariance matrix $\hat{\mathbf{e}}_1$, $\hat{\mathbf{e}}_2$ for three demonstrations at times $k = 10$, 30, and 44; (b) observed parameters for feature 1, $\mathbf{r}_1^{(1)}(t_{k+1})$, $\mathbf{r}_2^{(1)}(t_{k+1})$, $\mathbf{r}_3^{(1)}(t_{k+1})$. The vector $\mathbf{r}^{(1)}(t_{k+1}) - \mathbf{r}^{(1),\,\text{ref}}(t_{k+1})$ is required to lie in the region bounded by $\boldsymbol{\eta}_{\text{min}}$ and $\boldsymbol{\eta}_{\text{max}}$.

The matrix of the eigenvectors $\mathbf{E}_r(t_k)$ rotates the observed image feature vectors along the principal directions of the demonstrated motions. Thus, the observed motion is projected in a local reference frame, aligned with the instantaneous direction of the demonstrations. For instance, the observed parameters for feature 1 of the three demonstrations in the next time instant, $\mathbf{r}_1^{(1)}(t_{k+1}), \mathbf{r}_2^{(1)}(t_{k+1})$ and $\mathbf{r}_3^{(1)}(t_{k+1})$, are shown rotated in Figure 6.4b, with respect to the reference image feature parameters $\mathbf{r}^{(1),\,\text{ref}}(t_{k+1})$. The rotated observation vectors $\mathbf{r}_m^{(1)}(t_{k+1}) - \mathbf{r}^{(1),\,\text{ref}}(t_{k+1})$ for $m = 1,2,...,M$ define the boundaries of the demonstrated space at the time instant t_{k+1}, which corresponds to the hatched section in Figure 6.4b. The inner and outer bounds of the demonstrated envelope are calculated according to

$$
\begin{cases}
\boldsymbol{\eta}_{\min}(t_{k+1}) = \min_{m=1,2,...,M} \left\{ \mathbf{E}_r(t_{k+1}) \left(\mathbf{r}_m^{(1)}(t_{k+1}) - \mathbf{r}^{(1),\,\text{ref}}(t_{k+1}) \right) \right\} \\
\boldsymbol{\eta}_{\max}(t_{k+1}) = \max_{m=1,2,...,M} \left\{ \mathbf{E}_r(t_{k+1}) \left(\mathbf{r}_m^{(1)}(t_{k+1}) - \mathbf{r}^{(1),\,\text{ref}}(t_{k+1}) \right) \right\}
\end{cases}.
$$

$$(6.17)$$

The minimum and maximum operations in (6.17) are performed separately for the horizontal and vertical image coordinates, so that the bounds $\boldsymbol{\eta}_{\min} = \left[\eta_{\min}^{\hat{e}_1} \eta_{\min}^{\hat{e}_2} \right]$ and $\boldsymbol{\eta}_{\max} = \left[\eta_{\max}^{\hat{e}_1} \eta_{\max}^{\hat{e}_2} \right]$ represent 2×1 vectors (see Figure 6.4b).

Suppose there exists an unknown vector $\mathbf{r}^{(1)}(t_{k+1})$ related to the image feature parameters in the next time instance. The vector $\mathbf{r}^{(1)}(t_{k+1}) - \mathbf{r}^{(1),\,\text{ref}}(t_{k+1})$, and its coordinate transformation when rotated in the instantaneous demonstrated direction, is denoted by

$$
\boldsymbol{\eta}(t_{k+1}) = \mathbf{E}_r(t_{k+1}) \left(\mathbf{r}^{(1)}(t_{k+1}) - \mathbf{r}^{(1),\,\text{ref}}(t_{k+1}) \right). \tag{6.18}
$$

Then the following constraint ensures that each dimension of $\boldsymbol{\eta}(t_{k+1})$ is bounded within the demonstrated envelope:

$$
\begin{cases}
\eta_{\min}^{\hat{e}_1}(t_{k+1}) \leq \eta^{\hat{e}_1}(t_{k+1}) \leq \eta_{\max}^{\hat{e}_1}(t_{k+1}) \\
\eta_{\min}^{\hat{e}_2}(t_{k+1}) \leq \eta^{\hat{e}_2}(t_{k+1}) \leq \eta_{\max}^{\hat{e}_2}(t_{k+1})
\end{cases}. \tag{6.19}
$$

The inequalities in (6.19) can be converted to equalities by introducing non-negative excess (or surplus) and slack variables. In other words, by adding a vector of excess variables $\boldsymbol{\eta}_e$ to $\boldsymbol{\eta}$, and subtracting a vector of slack variables $\boldsymbol{\eta}_s$ from $\boldsymbol{\eta}$, the constraints (6.19) can be represented with the following linear equalities:

$$\begin{cases} \boldsymbol{\eta}(t_{k+1}) + \boldsymbol{\eta_e}(t_{k+1}) = \boldsymbol{\eta}_{\max}(t_{k+1}) \\ \boldsymbol{\eta}(t_{k+1}) - \boldsymbol{\eta_s}(t_{k+1}) = \boldsymbol{\eta}_{\min}(t_{k+1}) \end{cases}. \tag{6.20}$$

Constraint 3. This constraint ensures that the image feature trajectories stay in the FoV of the camera. Therefore, if the image limits in the horizontal direction are denoted as $r^{u,\,\min}$ and $r^{u,\,\max}$, and the vertical image limits are denoted by $r^{v,\,\min}$ and $r^{v,\,\max}$, then the coordinates of each feature point should stay bounded within the image limits, that is, the following set of inequalities should hold:

$$\begin{cases} r^{u,\,\min} \le r^{(n),u}(t_{k+1}) \le r^{u,\,\max} \\ r^{v,\,\min} \le r^{(n),v}(t_{k+1}) \le r^{v,\,\max} \end{cases}, \text{ for } n = 1,2,...,N. \tag{6.21}$$

By adding excess and slack variables, the constraints in (6.21) are rewritten as

$$\begin{cases} \mathbf{r}^{(n)}(t_{k+1}) + \mathbf{r_e}(t_{k+1}) = \mathbf{r}_{\max} \\ \mathbf{r}^{(n)}(t_{k+1}) - \mathbf{r_s}(t_{k+1}) = \mathbf{r}_{\min} \end{cases}, \text{ for } n = 1,2,...,N. \tag{6.22}$$

6.4.5.2 Cartesian Space Constraints

These constraints apply to the object position and velocity in the Cartesian space.

Constraint 1. The relationship between the Cartesian position of the object with respect to the camera frame and the translational (linear) velocity is

$$\frac{d}{dt}\mathbf{P}_o^c = \boldsymbol{v}_o^c(t). \tag{6.23}$$

In discrete form, (6.23) is represented as

$$\mathbf{P}_o^c(t_{k+1}) = \mathbf{P}_o^c(t_k) + \boldsymbol{v}_o^c(t_k)\Delta t_k. \tag{6.24}$$

Constraint 2. A constraint is introduced to guarantee that the Cartesian trajectory of the object stays within the envelope of the demonstrated motions in the Cartesian space. This constraint will prevent potential collisions of the object with the surrounding objects in the scene. Note that it is assumed that the demonstrated space defined by the envelope of the demonstrated Cartesian motions is free of obstacles, that is, all the points within the demonstrated envelope are considered safe.

Similar to the image-based Constraint 2 developed in (6.18)–(6.20), the inner and outer bounds of the demonstrations are found by utilizing the principal directions of the covariance matrix of the demonstrated Cartesian trajectories, i.e.,

$$\begin{cases} \boldsymbol{\mu}_{\min}(t_{k+1}) = \min_{m=1,2,\dots,M} \left\{ \mathbf{E}_\mathrm{p}(t_{k+1}) \left(\mathbf{P}_o^{c,m}(t_{k+1}) - \mathbf{P}_o^{c,\mathrm{ref}}(t_{k+1}) \right) \right\} \\ \boldsymbol{\mu}_{\max}(t_{k+1}) = \max_{m=1,2,\dots,M} \left\{ \mathbf{E}_\mathrm{p}(t_{k+1}) \left(\mathbf{P}_o^{c,m}(t_{k+1}) - \mathbf{P}_o^{c,\mathrm{ref}}(t_{k+1}) \right) \right\} \end{cases}.$$

$$(6.25)$$

For an unknown vector $\mathbf{P}_o^c(t_{k+1})$ associated with the position of the object of interest in the next time instant, the distance vector to the reference object position $\mathbf{P}_o^{c,\mathrm{ref}}(t_{k+1})$, rotated by the instantaneous eigenvector matrix, is given by

$$\boldsymbol{\mu}(t_{k+1}) = \mathbf{E}_\mathrm{p}(t_{k+1}) \left(\mathbf{P}_o^c(t_{k+1}) - \mathbf{P}_o^{c,\mathrm{ref}}(t_{k+1}) \right). \qquad (6.26)$$

The components of $\boldsymbol{\mu}(t_{k+1})$ are constrained to lie within the bounds of the demonstrated envelope via the following inequality:

$$\mu_{\min}^d(t_{k+1}) \le \mu^d(t_{k+1}) \le \mu_{\max}^d(t_{k+1}), \text{ for } d = 1,2,3, \qquad (6.27)$$

where d denotes the dimensionality of the vectors. By introducing excess $\boldsymbol{\mu}_\mathrm{e}$ and slack $\boldsymbol{\mu}_\mathrm{s}$ variables, the constraint can be represented in the form of equalities

$$\begin{cases} \boldsymbol{\mu}(t_{k+1}) + \boldsymbol{\mu}_\mathrm{e}(t_{k+1}) = \boldsymbol{\mu}_{\max}(t_{k+1}) \\ \boldsymbol{\mu}(t_{k+1}) - \boldsymbol{\mu}_\mathrm{s}(t_{k+1}) = \boldsymbol{\mu}_{\min}(t_{k+1}) \end{cases}. \qquad (6.28)$$

Constraint 3. Another constraint is established for the velocity of the object $\mathbf{v}_o$, which is to be bounded between certain minimum and maximum values,

$$v_{\min}^d \le v_o^d(t_k) \le v_{\min}^d, \text{ for } d = 1,2,\dots,6, \qquad (6.29)$$

where d is also used for indexing the dimensions of the velocity vector.

By plugging excess and slack variables in (6.29), the following set of equations is obtained:

$$\begin{cases} \mathbf{v}_o(t_k) + \mathbf{v}_\mathrm{e}(t_k) = v_{\max} \\ \mathbf{v}_o(t_k) - \mathbf{v}_\mathrm{s}(t_k) = v_{\min} \end{cases}. \qquad (6.30)$$

The values of maximal and minimal velocities, $\mathbf{v}_{\max}$ and $\mathbf{v}_{\min}$, could be associated with the extreme values of the velocities that can be exerted by the robot's end point during the object's manipulation. Therefore, this constraint can also be categorized into the robot kinematic constraints, presented next.

6.4.5.3 Robot Manipulator Constraints

This set of constraints establishes the relationship between the robot joint angles and the output variable of the optimization model (i.e., the velocity of the target object $\mathbf{v}_o^c$), and introduces a constraint regarding the limitations of the robot's joints.

Constraint 1. The first constraint relates the robot joint variables to the object's velocity. It is assumed that the object is grasped in the robot's gripper (Figure 6.3), with the velocity transformation between the object frame expressed in the robot base frame $\mathbf{v}_o^b = \left(\mathbf{v}_o^b, \boldsymbol{\omega}_o^b\right)$ and robot's endpoint frame $\mathbf{v}_e^b$ given by

$$\begin{cases} \mathbf{v}_o^b = \mathbf{v}_e^b + \boldsymbol{\omega}_e^b \times \mathbf{P}_{e,o}^b = \mathbf{v}_e^b - S\left(\mathbf{R}_e^b \mathbf{P}_o^e\right)\boldsymbol{\omega}_e^b \\ \boldsymbol{\omega}_o^b = \boldsymbol{\omega}_e^b \end{cases}. \tag{6.31}$$

In (6.31), the notation $\mathbf{P}_{e,o}^b$ is used for the position vector of the object with respect to the end point expressed relative to the robot base frame, whereas $S(\cdot)$ denotes a skew-symmetric matrix, which for an arbitrary vector $\mathbf{a} = \left[a_x, a_y, a_z\right]$ is defined as

$$S(\mathbf{a}) = \begin{bmatrix} 0 & -a_z & a_y \\ a_z & 0 & -a_x \\ -a_y & a_x & 0 \end{bmatrix}. \tag{6.32}$$

The differential kinematic equation of the robot is given with (6.2). Hence, the relationship between the joint variables and the object velocity in camera frame is obtained using (6.31) and (6.2),

$$\begin{aligned} \dot{\mathbf{q}}(t) &= \mathbf{J}^\dagger(\mathbf{q}(t))\mathbf{v}_e^b(t) \\ &= \mathbf{J}^\dagger(\mathbf{q}(t)) \begin{bmatrix} \mathbf{I}_{3\times3} & -S\left(\mathbf{R}_e^b(t)\mathbf{P}_o^e\right) \\ \mathbf{0}_{3\times3} & \mathbf{I}_{3\times3} \end{bmatrix}^{-1} \mathbf{v}_o^b(t) \\ &= \mathbf{J}^\dagger(\mathbf{q}(t)) \begin{bmatrix} \mathbf{I}_{3\times3} & -S\left(\mathbf{R}_e^b(t)\mathbf{P}_o^e\right) \\ \mathbf{0}_{3\times3} & \mathbf{I}_{3\times3} \end{bmatrix}^{-1} \begin{bmatrix} \mathbf{R}_c^b & \mathbf{0}_{3\times3} \\ \mathbf{0}_{3\times3} & \mathbf{R}_c^b \end{bmatrix} \mathbf{v}_o^c(t), \end{aligned} \tag{6.33}$$

where $\mathbf{I}_{3\times3}$ and $\mathbf{0}_{3\times3}$ are 3×3 identity and zeroes matrices, respectively, and $\mathbf{J}^\dagger(\mathbf{q})$ denotes the pseudoinverse of the robot Jacobian matrix of size $N_q \times 6$. At the time t_k, (6.33) can be represented in a discrete form

$$\begin{aligned} \mathbf{q}(t_{k+1}) = \mathbf{q}(t_k) + \mathbf{J}^\dagger(\mathbf{q}(t_k)) \begin{bmatrix} \mathbf{I}_{3\times3} & -S\left(\mathbf{R}_e^b(t_k)\mathbf{P}_o^e\right) \\ \mathbf{0}_{3\times3} & \mathbf{I}_{3\times3} \end{bmatrix}^{-1} \\ \begin{bmatrix} \mathbf{R}_c^b & \mathbf{0}_{3\times3} \\ \mathbf{0}_{3\times3} & \mathbf{R}_c^b \end{bmatrix} \mathbf{v}_o^c(t_k)\Delta t_k. \end{aligned} \tag{6.34}$$

The rotation matrix of robot's end point in base frame $\mathbf{R}_e^b(t_k)$ is obtained by using the robot's forward kinematics. The rotation matrix of the camera frame in robot base frame $\mathbf{R}_c^b$ is found from the calibration of the camera. This matrix is constant, since both the robot's base frame and the camera frame are fixed. Position of the object frame with respect to the end-point frame $\mathbf{P}_o^e$ is also time-independent, and it is obtained by measurements with a coordinate-measuring machine.

Constraint 2. A constraint ensuring that the robot joint variables are within the robot workspace limits is defined as

$$q_{\min}^d \le q^d(t_{k+1}) \le q_{\max}^d \text{ for } d = 1,2,...,N_q, \tag{6.35}$$

where $\mathbf{q}_{\min}$ and $\mathbf{q}_{\max}$ stand for the vectors of minimal and maximal realizable values for the joint angles. To represent (6.35) in the form of equalities, excess and slack variables are introduced yielding

$$\begin{cases} \mathbf{q}(t_{k+1}) + \mathbf{q}_e(t_{k+1}) = \mathbf{q}_{\max} \\ \mathbf{q}(t_{k+1}) - \mathbf{q}_s(t_{k+1}) = \mathbf{q}_{\min} \end{cases}. \tag{6.36}$$

6.4.6 Optimization Model

The overall second-order optimization model is reported in this section in its full form. The model is rewritten in terms of the unknown (decision) variables, and the known variables and parameters, at each instance of the optimization procedure.

Recall that the objective function is defined in (6.14). The model constraints include the following: the linear constraints defining the relations between the variables given with (6.16), (6.18), (6.24), (6.26), and (6.34); the linear constraints obtained from inequalities by introducing excess and slack variables given with (6.20), (6.22), (6.28), (6.30), and (6.36); and the conic constraints given with (6.12) and (6.13). Also, non-negativity constraints for the components of all excess and slack variables are included in the model.

Auxiliary variables are introduced that denote the difference between the image parameters and the reference image features for the feature n at the next time instant, $\bar{\mathbf{r}}^{(n)}(t_{k+1}) = \mathbf{r}^{(n)}(t_{k+1}) - \mathbf{r}^{(n),\text{ref}}(t_{k+1})$. Accordingly, the corresponding image feature vector is obtained by stacking the variables for each image feature, $\bar{\boldsymbol{\xi}} = \begin{bmatrix} \bar{\mathbf{r}}^{(1)T} & \bar{\mathbf{r}}^{(2)T} & \cdots & \bar{\mathbf{r}}^{(n)T} \end{bmatrix}^T$. Thus, subtracting the term $\boldsymbol{\xi}^{\text{ref}}(t_{k+1})$ on both sides of (6.16) results in

$$\boldsymbol{\xi}(t_{k+1}) - \boldsymbol{\xi}^{\text{ref}}(t_{k+1}) = \boldsymbol{\xi}(t_k) - \boldsymbol{\xi}^{\text{ref}}(t_{k+1}) + \mathbf{L}(t_k)\mathbf{v}_o^c(t_k)\Delta t_k. \tag{6.37}$$

With rearranging the given equation, it can be rewritten in terms of the unknown variables in the model $\bar{\xi}(t_{k+1})$ and $\mathbf{v}_o^c(t_k)$, and the known variables at the time instant t_k, that is, $\xi(t_k)$, $\mathbf{L}(t_k)$, $\xi^{\mathrm{ref}}(t_{k+1})$, and Δt_k

$$\begin{bmatrix} \mathbf{I} & -\mathbf{L}(t_k)\Delta t_k \end{bmatrix} \begin{bmatrix} \bar{\xi}(t_{k+1}) \\ \mathbf{v}_o^c(t_k) \end{bmatrix} = \xi(t_k) - \xi^{\mathrm{ref}}(t_{k+1}). \tag{6.38}$$

Similarly, (6.18) is rewritten in terms of the introduced variables

$$\begin{bmatrix} \mathbf{I} & -\mathbf{E}_r^{(n)}(t_{k+1}) \end{bmatrix} \begin{bmatrix} \boldsymbol{\eta}^{(n)}(t_{k+1}) \\ \bar{\mathbf{r}}^{(n)}(t_{k+1}) \end{bmatrix} = \mathbf{0}, \quad \text{for } n = 1,2,...,N. \tag{6.39}$$

Analogous to (6.38), by introducing an auxiliary variable $\bar{\mathbf{P}}_o^c(t_{k+1}) = \mathbf{P}_o^c(t_{k+1}) - \mathbf{P}_o^{c,\mathrm{ref}}(t_{k+1})$, and subtracting $\mathbf{P}_o^{c,\mathrm{ref}}(t_{k+1})$ from both sides of (6.24), it becomes

$$\begin{bmatrix} \mathbf{I} & -\Delta t_k \mathbf{I} \end{bmatrix} \begin{bmatrix} \bar{\mathbf{P}}_o^c(t_{k+1}) \\ \boldsymbol{v}_o^c(t_k) \end{bmatrix} = \mathbf{P}_o^c(t_k) - \mathbf{P}_o^{c,\mathrm{ref}}(t_{k+1}), \tag{6.40}$$

whereas (6.26) can also be rewritten in terms of the unknown variables

$$\begin{bmatrix} \mathbf{I} & -\mathbf{E}_P(t_{k+1}) \end{bmatrix} \begin{bmatrix} \boldsymbol{\mu}(t_{k+1}) \\ \bar{\mathbf{P}}_o^c(t_{k+1}) \end{bmatrix} = \mathbf{0}. \tag{6.41}$$

In these equations, the variables $\mathbf{P}_o^c(t_k)$, $\mathbf{P}_o^{c,\mathrm{ref}}(t_{k+1})$, and $\mathbf{E}_P(t_{k+1})$ are known.

Similarly, (6.34) can be rewritten as

$$\begin{bmatrix} \mathbf{I} & -\mathbf{J}^\dagger(\mathbf{q}(t_k))\Delta t_k \begin{bmatrix} \mathbf{I}_{3\times3} & -S(\mathbf{R}_e^b\mathbf{P}_o^e) \\ \mathbf{0}_{3\times3} & \mathbf{I}_{3\times3} \end{bmatrix}^{-1} \begin{bmatrix} \mathbf{R}_c^b & \mathbf{0}_{3\times3} \\ \mathbf{0}_{3\times3} & \mathbf{R}_c^b \end{bmatrix} \end{bmatrix}$$
$$\begin{bmatrix} \mathbf{q}(t_{k+1}) \\ \mathbf{v}_o^c(t_k) \end{bmatrix} = \mathbf{q}(t_k). \tag{6.42}$$

The linear constraints related to the excess and slack variables in (6.20), (6.22), (6.28), (6.30), and (6.36) can be represented in a similar form as

$$\begin{bmatrix} \mathbf{I} & \mathbf{I} \end{bmatrix} \begin{bmatrix} \boldsymbol{\eta}(t_{k+1}) \\ \boldsymbol{\eta}_e(t_{k+1}) \end{bmatrix} = \boldsymbol{\eta}_{\max}(t_{k+1})$$
$$\begin{bmatrix} \mathbf{I} & -\mathbf{I} \end{bmatrix} \begin{bmatrix} \boldsymbol{\eta}(t_{k+1}) \\ \boldsymbol{\eta}_s(t_{k+1}) \end{bmatrix} = \boldsymbol{\eta}_{\min}(t_{k+1})$$
$$\tag{6.43}$$

$$[\mathbf{I} \quad \mathbf{I}] \begin{bmatrix} \bar{\mathbf{r}}^{(n)}(t_{k+1}) \\ \mathbf{r_e}(t_{k+1}) \end{bmatrix} = \mathbf{r}_{\max} - \mathbf{r}^{(n),\mathrm{ref}}(t_{k+1})$$

$$[\mathbf{I} \quad -\mathbf{I}] \begin{bmatrix} \bar{\mathbf{r}}^{(n)}(t_{k+1}) \\ \mathbf{r_s}(t_{k+1}) \end{bmatrix} = \mathbf{r}_{\min} - \mathbf{r}^{(n),\mathrm{ref}}(t_{k+1})$$
, for $n = 1, 2, \ldots, N,$

$$\tag{6.44}$$

$$[\mathbf{I} \quad \mathbf{I}] \begin{bmatrix} \boldsymbol{\mu}(t_{k+1}) \\ \boldsymbol{\mu_e}(t_{k+1}) \end{bmatrix} = \boldsymbol{\mu}_{\max}(t_{k+1})$$

$$[\mathbf{I} \quad -\mathbf{I}] \begin{bmatrix} \boldsymbol{\mu}(t_{k+1}) \\ \boldsymbol{\mu_s}(t_{k+1}) \end{bmatrix} = \boldsymbol{\mu}_{\min}(t_{k+1})$$

$$\tag{6.45}$$

$$[\mathbf{I} \quad \mathbf{I}] \begin{bmatrix} \mathbf{v}_o(t_k) \\ \mathbf{v_e}(t_k) \end{bmatrix} = \mathbf{v}_{\max}$$

$$[\mathbf{I} \quad -\mathbf{I}] \begin{bmatrix} \mathbf{v}_o(t_k) \\ \mathbf{v_s}(t_k) \end{bmatrix} = \mathbf{v}_{\min}$$

$$\tag{6.46}$$

$$[\mathbf{I} \quad \mathbf{I}] \begin{bmatrix} \mathbf{q}(t_{k+1}) \\ \mathbf{q_e}(t_{k+1}) \end{bmatrix} = \mathbf{q}_{\max}$$

$$[\mathbf{I} \quad -\mathbf{I}] \begin{bmatrix} \mathbf{q}(t_{k+1}) \\ \mathbf{q_s}(t_{k+1}) \end{bmatrix} = \mathbf{q}_{\min}$$

$$\tag{6.47}$$

The conic constraint (6.12) can also be rewritten in a similar fashion as

$$\tau_n \le \left\| \bar{\mathbf{r}}^{(n)}(t_{k+1}) \right\|, \quad \text{for } n = 1, 2, \ldots, N. \tag{6.48}$$

The other conic constraints (6.13) can be written as

$$\tau_{\mathbf{v}} \le \left\| \bar{\mathbf{v}}_o^c(t_k) \right\|$$
$$\tau_{\boldsymbol{\omega}} \le \left\| \bar{\boldsymbol{\omega}}_o^c(t_k) \right\|, \tag{6.49}$$

where auxiliary variables for the linear and angular velocity are introduced as $\bar{\mathbf{v}}_o^c(t_k) = \mathbf{v}_o^c(t_k) - \mathbf{v}_o^{c,\mathrm{ref}}(t_k)$ and $\bar{\boldsymbol{\omega}}_o^c(t_k) = \boldsymbol{\omega}_o^c(t_k) - \boldsymbol{\omega}_o^{c,\mathrm{ref}}(t_k)$, respectively, that is,

$$[\mathbf{I} \quad -\mathbf{I}] \begin{bmatrix} \mathbf{v}_o^c(t_k) \\ \bar{\mathbf{v}}_o^c(t_k) \end{bmatrix} = \mathbf{v}_o^{c,\mathrm{ref}}(t_k).$$

$$[\mathbf{I} \quad -\mathbf{I}] \begin{bmatrix} \boldsymbol{\omega}_o^c(t_k) \\ \bar{\boldsymbol{\omega}}_o^c(t_k) \end{bmatrix} = \boldsymbol{\omega}_o^{c,\mathrm{ref}}(t_k)$$

$$\tag{6.50}$$

The non-negativity constraints for the excess and slack variables yield

$$\boldsymbol{\eta_e}(t_{k+1}), \boldsymbol{\eta_s}(t_{k+1}), \mathbf{r_e}(t_{k+1}), \mathbf{r_s}(t_{k+1}), \boldsymbol{\mu_e}(t_{k+1}), \boldsymbol{\mu_s}(t_{k+1}) \geq 0$$
$$\mathbf{v_e}(t_{k+1}), \mathbf{v_s}(t_{k+1}), \mathbf{q_e}(t_{k+1}), \mathbf{q_s}(t_{k+1}) \geq 0. \tag{6.51}$$

To summarize, the optimization variable $\mathbf{z}$ in (6.11) at the time instant t_k is formed by concatenation of the variables from the conic constraints (6.48) and (6.49)

$$\mathbf{z}_c(t_k) = \left[\tau_1, \bar{\mathbf{r}}^{(1)}(t_{k+1}), \ldots, \tau_N, \bar{\mathbf{r}}^{(n)}(t_{k+1}), \tau_\upsilon, \bar{\mathbf{v}}_o^c(t_k), \tau_\omega, \bar{\boldsymbol{\omega}}_o^c(t_k) \right],$$
$$\tag{6.52}$$

and the variables from the linear constraints

$$\mathbf{z}_l(t_k) = [\boldsymbol{\eta}(t_{k+1}), \boldsymbol{\mu}(t_{k+1}), \bar{\mathbf{P}}_o^c(t_{k+1}), \mathbf{q}(t_{k+1}), \boldsymbol{\eta_e}(t_{k+1}), \boldsymbol{\eta_s}(t_{k+1}), \boldsymbol{\xi_e}(t_{k+1}),$$
$$\boldsymbol{\xi_s}(t_{k+1}), \boldsymbol{\mu_e}(t_{k+1}), \boldsymbol{\mu_s}(t_{k+1}), \mathbf{v_e}(t_k), \mathbf{v_s}(t_k), \mathbf{q_e}(t_{k+1}), \mathbf{q_s}(t_{k+1})]$$
$$\tag{6.53}$$

that is, $\mathbf{z}(t_k) = \left[\mathbf{z}_c(t_k)^T \; \mathbf{z}_l(t_k)^T \right]^T$. The size of the component $\mathbf{z}_c(t_k)$ is $3N + 8$, whereas the size of the component $\mathbf{z}_l(t_k)$ is $10N + 3N_q + 24$.

From the cost function defined in (6.14)

$$\text{minimize} \left\{ \sum_{n=1}^{N} \alpha_n \tau_n + \alpha_\upsilon \tau_\upsilon + \alpha_\omega \tau_\omega \right\}$$

the part of the vector $\mathbf{c}$ in (6.11) that corresponds to the $\mathbf{z}_c(t_k)$ is

$$\mathbf{c}_c(t_k) = [\alpha_1 \; \mathbf{0}_{1\times2} \; \cdots \; \alpha_N \; \mathbf{0}_{1\times2} \; \alpha_\upsilon \; \mathbf{0}_{1\times3} \; \alpha_\omega \; \mathbf{0}_{1\times3}]^T, \tag{6.54}$$

whereas the part $\mathbf{c}_l(t_k)$ corresponding to $\mathbf{z}_l(t_k)$ is all zeros, since those variables are not used in the cost function.

The known parameters for the optimization model at time t_k are as follows: Δt_k, $\boldsymbol{\xi}(t_k)$, $\boldsymbol{\xi}^{\text{ref}}(t_{k+1})$, $\mathbf{L}(t_k)$, $\mathbf{E}_r^{(n)}(t_{k+1})$, $\mathbf{P}_o^c(t_k)$, $\mathbf{P}_o^{c,\text{ref}}(t_{k+1})$, $\mathbf{E}_P(t_{k+1})$, $\mathbf{q}(t_k)$, $\mathbf{R}_e^b(t_k)$, $\mathbf{J}^\dagger(\mathbf{q}(t_k))$, $\mathbf{v}_o^{c,\text{ref}}(t_k)$, $\boldsymbol{\eta}_{\min}(t_{k+1})$, $\boldsymbol{\eta}_{\max}(t_{k+1})$, $\boldsymbol{\mu}_{\min}(t_{k+1})$, $\boldsymbol{\mu}_{\max}(t_{k+1})$, and the time-independent parameters: $\mathbf{r}_{\min}$, $\mathbf{r}_{\max}$, $\mathbf{v}_{\min}$, $\mathbf{v}_{\max}$, $\mathbf{q}_{\min}$, $\mathbf{q}_{\max}$, $\mathbf{R}_c^b$, and $\mathbf{P}_o^e$.

The problem can be solved as second-order conic minimization because the relations among the variables and the parameters are formulated as linear and conic constraints. A series of the presented minimization procedure is performed at each time instant t_k. The important unknown variables of the model at times t_k are the object velocity $\mathbf{v}_o^c(t_k)$ and the locations of the features points at the next time instant $\boldsymbol{\xi}(t_{k+1})$. By tuning the weighting coefficients α's in the objective function, the results can be adjusted to follow more closely the reference image trajectories, or the reference object velocities, as explained in the sequel.

6.5 Robust Image-Based Tracking Control

To follow the image feature trajectories $\xi(t_k)$ for $k = 1, 2, \ldots$ generated from the optimization model, an image-based visual tracker is employed. The control ensures that the errors between the measured feature parameters $\bar{\xi}$ and the desired feature parameters ξ, that is, $\mathbf{e}(t) = \bar{\xi}(t) - \xi(t)$, are driven to zero for $t \in (0, \infty)$. By taking derivative of the error and using (6.15), a relation between the features error and the object velocity is established as follows:

$$\dot{\mathbf{e}}(t) = \dot{\bar{\xi}}(t) - \dot{\xi}(t) = \mathbf{L}(t)\mathbf{v}_o^c(t) - \dot{\xi}(t). \tag{6.55}$$

A controller with exponential decoupled decrease of the error $\dot{\mathbf{e}} = -\lambda\mathbf{e}$ is a common choice in solving visual servoing problems, thus it is adopted here as well. Therefore, (6.55) yields

$$\mathbf{v}_o^c(t) = -\lambda\hat{\mathbf{L}}^{\dagger}(t)\mathbf{e}(t) + \hat{\mathbf{L}}^{\dagger}(t)\dot{\xi}(t), \tag{6.56}$$

where $\hat{\mathbf{L}}^{\dagger}(t)$ is an approximation of the pseudoinverse of the image Jacobian matrix $\mathbf{L}(t)$. The applied control law warrants that when the error between the measured and the followed feature parameters is small, the velocity of the object will follow closely the desired velocity generated by the optimization model. Note that the image Jacobian matrix $\mathbf{L}(t)$ requires information that is not directly available from the image measurements, for example, partial pose estimation of the object. Therefore, an approximation of the matrix is usually used in visual servoing tasks, with different models for the approximation reported in the literature (Chaumette and Hutchinson, 2007).

Regarding the stability of the presented control scheme, it is well known that local asymptotic stability is guaranteed in the neighborhood of $\mathbf{e} = \mathbf{0}$ if $\hat{\mathbf{L}}^{\dagger}\mathbf{L}$ is positive definite (Chaumette and Hutchinson, 2006). Global asymptotic stability of this type of control cannot be achieved, because the matrix $\hat{\mathbf{L}}^{\dagger}$ has a nonzero null space. However, in the neighborhood of the desired pose, the control scheme is free of local minima and the convergence is guaranteed (Chaumette and Hutchinson, 2006). These properties of the IBVS control render its suitability for tracking tasks, such as the problem at hand. That is, when the features selection and system calibration are reasonably performed so that $\hat{\mathbf{L}}^{\dagger}\mathbf{L}$ is positive definite, then the errors between the current and desired feature parameters will converge to zero along the tracked trajectory. Although the region of local asymptotic stability is difficult to be calculated theoretically, it has been shown that, in practice, it can be quite large (Chaumette and Hutchinson, 2006).

For calculations of the robot joint angles for the task execution, the robot Jacobian matrix is combined with the image Jacobian matrix into a feature Jacobian matrix $\mathbf{J}_s \in \mathbb{R}^{2n \times N_q}$ (Chaumette and Hutchinson, 2007)

$$\mathbf{J}_s(\mathbf{q},t) = \mathbf{L}(t) \begin{bmatrix} \left(\mathbf{R}_c^b\right)^T & \mathbf{0}_{3 \times 3} \\ \mathbf{0}_{3 \times 3} & \left(\mathbf{R}_c^b\right)^T \end{bmatrix} \begin{bmatrix} \mathbf{I}_{3 \times 3} & -S\left(\mathbf{R}_e^b(t)\mathbf{P}_o^e\right) \\ \mathbf{0}_{3 \times 3} & \mathbf{I}_{3 \times 3} \end{bmatrix} \mathbf{J}(\mathbf{q}(t)),$$

(6.57)

which provides a direct mapping between the image feature parameters and the time change of the joint angle variables, that is, $\dot{\boldsymbol{\xi}}(t) = \mathbf{J}_s(\mathbf{q},t)\dot{\mathbf{q}}(t)$. The joint angles of the robot are updated based on (6.33) and (6.56), i.e.,

$$\dot{\mathbf{q}}(t) = -\lambda \hat{\mathbf{J}}_s^\dagger(\mathbf{q},t)\mathbf{e}(t) + \hat{\mathbf{J}}_s^\dagger(\mathbf{q},t)\dot{\boldsymbol{\xi}}(t),$$

(6.58)

which in discrete form becomes

$$\mathbf{q}(t_{k+1}) = \mathbf{q}(t_k) - \lambda \hat{\mathbf{J}}_s^\dagger(\mathbf{q},t_k)\mathbf{e}(t_k)\Delta t_k + \hat{\mathbf{J}}_s^\dagger(\mathbf{q},t_k)\dot{\boldsymbol{\xi}}(t_k)\Delta t_k.$$

(6.59)

6.5.1 Simulations

The presented approach is first evaluated through simulations in a virtual environment. The goal of the simulations is mainly to verify the developed task planning method in the image space via the presented optimization model. In the next section, the approach is subjected to more thorough evaluations through experiments with a robot in a real-world environment.

The simulations and the experiments were performed in the MATLAB environment on a 3-GHz quad-core CPU with 4-GB of RAM running under Windows 7 OS.

A simulated environment in MATLAB was created using functions from the visual servoing toolbox for MATLAB/Simulink (Cervera, 2003) and the Robotics toolbox for MATLAB (Corke, 2011). The conic optimization is solved by using the SeDuMi software package (Sturm, 1997).

A virtual camera model is created, corresponding to the 640 × 480 pixels Point Grey's Firefly®MV camera, introduced in Section 2.2. A virtual robot Puma 560 is adopted for execution of the generated reproduction strategy. Nonetheless, the focus of the simulations is on the task planning, rather than the task execution.

The simulations are carried out for two different types of demonstrated motions. Simulation 1 uses synthetic generated trajectories, whereas

Simulation 2 employs human-executed trajectories. The discrete sampling period for both simulations is set equal to $\Delta t_k = 0.1$ seconds, for $k = 1, 2, \ldots$.

Note that in Figures 6.5–6.14, the plots depicting the image plane of the camera are displayed with borders around them, in order to be easily differentiated from the time plots of the other variables, such as Cartesian positions or velocities. The image feature parameters in the figures are displayed in pixel coordinates $\mathbf{u}_m^{(n)}(t_k)$, rather than in spatial coordinates $\mathbf{r}_m^{(n)}(t_k)$, since it is more intuitive for presentation purposes. The demonstrated trajectories are depicted with thin solid lines (unless otherwise indicated). The initial states of the trajectories are indicated by square marks, while the final states are indicated by cross marks. In addition, the reference trajectories obtained from Kalman smoothing are represented by thick dashed lines, and the generalized trajectories obtained from the conic optimization are depicted by thick solid lines. The line styles are also described in the figure legends.

6.5.1.1 Simulation 1

This simulation case is designed for initial examination of the planning in a simplistic scenario with three synthetic trajectories of a moving object generated in MATLAB. The Cartesian positions of the object are shown in Figure 6.5a. For one of the demonstrated trajectories, the axes of the coordinate frame of the object are also shown in the plot. A total rotation of $60°$ around the y-axis has been applied to the object during the motion. The motions are characterized by constant velocities. Six planar features located on the object are observed by a stationary camera (i.e., $N = 6$). The projections of the feature points onto the image plane of the camera for the three demonstrations are displayed in Figure 6.5b with different line styles. The boundaries of the task in the Cartesian and image space can be easily inferred for the demonstrated motions. The simulation scenario is designed such that there is a violation of the FoV of the camera. Note that in the simulations the FoV of the camera is infinite, that is, negative values of the pixels can be handled.

Kalman smoothers are employed to find smooth average trajectories for each feature point $\mathbf{r}^{(n),\text{ref}}$, as well as to find the reference velocities of the object $\mathbf{v}_o^{\text{ref}}$. For initialization of the Kalman smoothers, the measurement and process noise covariance matrices are set as follows. For the image feature parameters corresponding to each feature point n: $\Sigma_{\mathbf{r}}^n = 10\mathbf{I}_{(4 \times 4)}$, $\mathbf{Y}_{\mathbf{r}}^n = 10\mathbf{I}_{(4M \times 4M)}$; for the object's pose: $\Sigma_{\mathbf{P}_o} = 0.1\mathbf{I}_{(6 \times 6)}$, $\mathbf{Y}_{\mathbf{P}_o} = 10\mathbf{I}_{(6M \times 6M)}$; and for the object's linear and angular velocities: $\Sigma_{\mathbf{v}_o} = 0.1\mathbf{I}_{(6 \times 6)}$, $\mathbf{Y}_{\mathbf{v}_o} = 10\mathbf{I}_{(6M \times 6M)}$. As noted before, the notation $\mathbf{I}_{(a \times a)}$ refers to an identity matrix of size a. The resulting reference trajectories for the image features

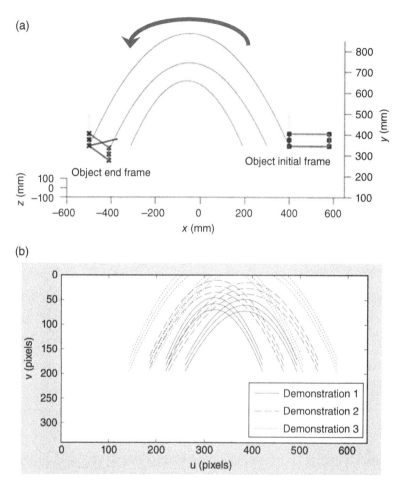

Figure 6.5 (a) Three demonstrated trajectories of the object in the Cartesian space. For one of the trajectories, the initial and ending coordinate frames of the object are shown, along with the six coplanar features; (b) projected demonstrated trajectories of the feature points onto the image plane of the camera; (c) reference image feature trajectories produced by Kalman smoothing and generalized trajectories produced by the optimization model; and (d) object velocities from the optimization model.

are depicted with dashed lines in Figure 6.5c. It can be noticed that for some of the feature points, a section of the reference trajectories is outside of the image boundaries.

The optimization model constraints (6.22) ensure that all image features stay within the camera FoV during the entire length of the motion. The image boundaries in this case are defined as ±10 pixels from the horizontal and vertical dimensions of the image sensor, that is,

(c)

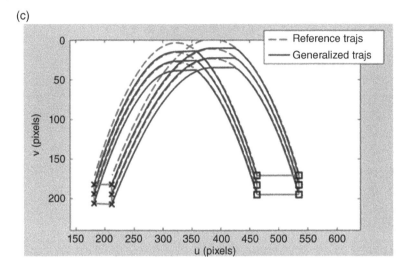

(d)

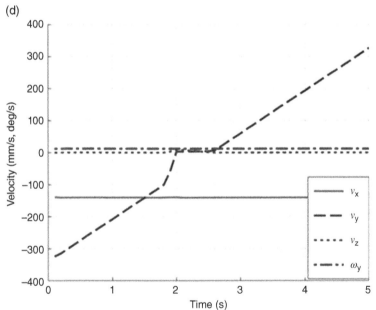

Figure 6.5 (*Continued*)

$$\begin{cases} u^{\max} = 630 \\ u^{\min} = 10 \end{cases}, \begin{cases} v^{\max} = 470 \\ v^{\min} = 10 \end{cases}. \tag{6.60}$$

Accordingly, a series of second-order conic optimization models described in Section 6.4.6 is run, with the reference trajectories from the Kalman smoothing algorithms used as inputs for the optimizations. The following values for the weighting coefficients were adopted: $\{\alpha_n\}_{n=1}^6 = 0.1$ and $\alpha_v = \alpha_\omega = 0.5$. The resulting image feature trajectories from the optimization are depicted with thick solid lines in Figure 6.5c, whereas the resulting velocity of the object is shown in Figure 6.5d. The angular velocities around the x- and z-axes are zero and therefore are not shown in the figure. It can be observed in Figure 6.5c that the image feature trajectories are confined within the FoV bounds defined in (6.60). For the image feature parameters that have values less than the specified boundary for the vertical coordinate, that is, $u^{\min} = 10$ pixels, the presented approach modified the plan to satisfy the image limits constraint.

6.5.1.2 Simulation 2

This simulation entails a case where an object is moved by a human demonstrator, while the poses of the object are recorded with an optical marker-based tracker. A total of five demonstrations are collected, displayed in Figure 6.6. The demonstrations involve translatory motions, whereas synthetic rotation of $60°$ around the y-axis has been added to all trajectories. It is assumed that the object has six point features and the model of the object is readily available. Different from Simulation 1, the demonstrated trajectories in Simulation 2 are characterized by the nonconsistency of human-demonstrated motions. The velocities of the demonstrated trajectories are shown in Figure 6.6b. The Kalman-smoothed velocities of the object are shown with the dashed lines in Figure 6.6b. Similarly, the smoothed image feature trajectories are given in Figure 6.6c. The same initialization parameters for the Kalman smoothers as in Simulation 1 were employed.

The resulting reference trajectories are then used for initialization of the optimization procedure. Similar to Simulation 1, the weighting coefficients are set to $\{\alpha_n\}_{n=1}^6 = 0.1$ and $\alpha_v = \alpha_\omega = 0.5$. The weighting scheme assigns higher weight to the importance of following the reference velocities, whereas the model constraints ensure that the generated feature trajectories in the image space are close to the reference trajectories and are within the bounds of the demonstrated task space. As a result, the generated object velocities $\mathbf{v}_o$ are almost identical to the reference velocities shown in Figure 6.6b. The generated image feature trajectories are shown with the thick lines in Figure 6.6c. The corresponding Cartesian object trajectory is shown in Figure 6.6d. The trajectory is within the bounds of the demonstrated task space, and reaching the final goal state of the task is accomplished. The simulation results of the robot's task execution are not presented, since a more proficient analysis is provided in the next section, involving real-world experiments.

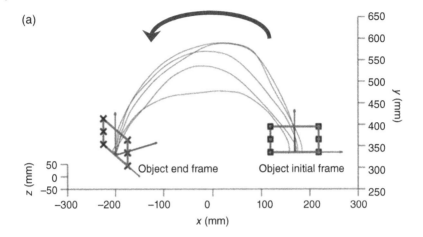

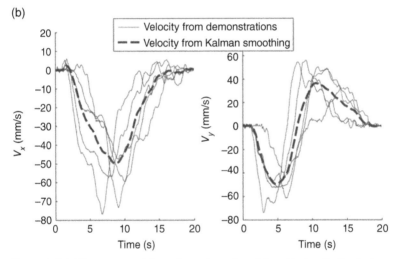

Figure 6.6 (a) Demonstrated Cartesian trajectories of the object, with the features points, and the initial and ending object frames; (b) demonstrated and reference linear velocities of the object for *x*- and *y*-coordinates of the motions; (c) reference image feature trajectories from the Kalman smoothing and the corresponding generalized trajectories from the optimization; and (d) the demonstrated and retrieved generalized object trajectories in the Cartesian space. The initial state and the ending state are depicted with square and cross marks, respectively.

6.5.2 Experiments

Experimental evaluation of the approach is conducted with a CRS A255 desktop robot and a Point Grey's Firefly MV camera (Section 2.2), both shown in Figure 6.7a. An object with five coplanar circular features shown

(c)

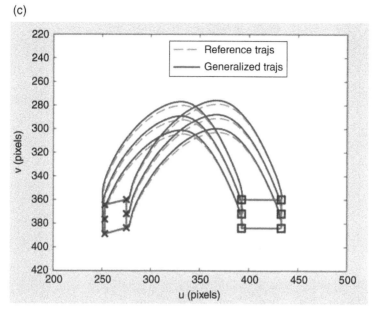

(d)

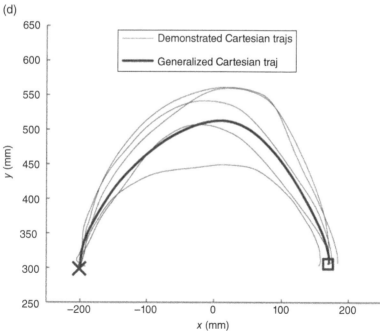

Figure 6.6 (*Continued*)

(a)

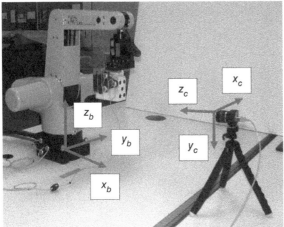

(b)

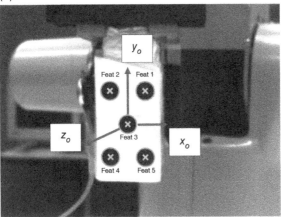

Figure 6.7 (a) Experimental setup showing the robot in the home configuration and the camera. The coordinate axes of the robot base frame and the camera frame are depicted; (b) the object with the coordinate frame axes and the features.

in Figure 6.7b is used for the experiments. The centroids of the circular features are depicted with cross markers in the figure, and also the circular features are circumscribed with solid lines. The coordinates of the centroids of the dots in the object frame are $(12, 22, 0)$, $(-12, 22, 0)$, $(0, 0, 0)$, $(-12, -22, 0)$, and $(12, -22, 0)$ millimeters. The demonstrations consist of manipulating the object in front of the camera.

The frame rate of the camera was set to 30 fps. A megapixel fixed focal length lens from Edmund Optics Inc. was used. The camera was calibrated by using the camera calibration toolbox for MATLAB® (Bouguet, 2010). The calibration procedure uses a set of 20 images of a planar checkerboard taken at different positions and orientations with respect to the camera. The camera parameters are estimated employing nonlinear optimization (based on iterative gradient descent) for minimizing the radial and tangential distortion in the images. The calibration estimated values of the intrinsic camera parameters are as follows: principal point coordinates $u_0 = 296.54$, $v_0 = 266.04$ pixels, focal length $f_C = 8.3$ millimeters, and the scaling factors $f_C k_u = 1395.92$, $f_C k_v = 1397.88$ pixels.

The CRS A255 robot is controlled through the QuaRC toolbox for open-architecture control (QuaRC, 2012). In fact, the CRS A255 robot was originally acquired as a closed-architecture system, meaning that pre-structured proportional–integral–derivative (PID) feedback controllers are employed for actuating the links' joints. For example, when a user commands the robot to move to a desired position, the built-in PID controllers calculate and send current to joint motors, in order to drive the robot to the desired position. That is, the closed-architecture controller prevents the users from accessing the joint motors and implementing advanced control algorithms. On the other hand, the QuaRC toolbox for open architecture, developed by Quanser (Markham, Ontario), allows direct control of the joint motors of the CRS robot. It uses Quanser Q8 data-acquisition board for reading the joint encoders and for communicating with the motor amplifiers at 1-millisecond intervals. In addition, QuaRC provides a library of custom Simulink blocks and S-functions developed for the A255 robot, thus enabling design of advanced control algorithms and real-time control of the robot motions in Simulink environment. The toolbox also contains a block for acquisition of images with Point Grey's FireflyMV camera.

The A255 robot has an anthropomorphic structure, that is, it is realized entirely by revolute joints, and by analogy to the human body, it consists of an arm and a wrist. The arm of the A255 manipulator endows three DoFs, whereas the wrist provides two DoFs. The fact that the wrist has only two DoFs prevents from exerting arbitrary orientation of the robot's end point, and subsequently an arbitrary orientation of the manipulated object cannot be achieved. Accordingly, the experiments are designed to avoid the shortcomings of the missing rotational DoF. In the first place, one must have in mind that human-demonstrated trajectories cannot be completely devoid of a certain rotational DoF, due to the stochastic character of human motion. Therefore, kinesthetic demonstrations are employed for the experiments (see Figure 6.8), meaning that while the object is grasped by the robot's gripper, the joint motors are set into passive mode,

Figure 6.8 Sequence of images from the kinesthetic demonstrations.

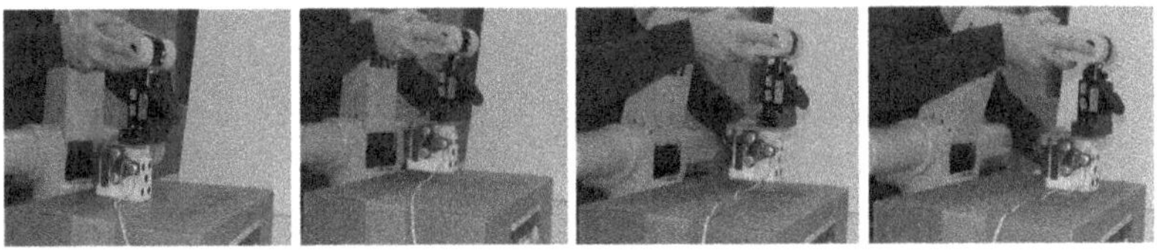

and the links are manually moved by a demonstrator (Hersch *et al.*, 2008; Kormushev *et al.*, 2011). Furthermore, the kinesthetic mode of demonstrations does not require solving the correspondence problem between the embodiments of the demonstrator and the robot. The drawbacks are related to the reduced naturalness of the motions, and also this demonstration mode can be challenging for robots with large number of DoFs.

Three sets of experiments are designed to evaluate the performance of the presented approach.

6.5.2.1 Experiment 1

The first experiment consists of moving an object along a trajectory of a pick-and-place task. A sequence of images from the kinesthetic demonstrations is displayed in Figure 6.8. The total number of demonstrated trajectories M is 5. Image plane coordinates of the centroids of the dots are considered as image feature parameters, denoted by $\boldsymbol{\xi} = \left[\mathbf{r}^{(1)^T}, \mathbf{r}^{(2)^T}, \mathbf{r}^{(3)^T}, \mathbf{r}^{(4)^T}, \mathbf{r}^{(5)^T} \right]^T$. The trajectories of the feature points in the image plane of the camera for one of the demonstrations are presented in Figure 6.9a. As explained, the initial and ending states are illustrated with square and cross marks, respectively.

Tracking the features in the acquired image sequences is based on the "dot tracker" method reported in the VISP package (Marchand *et al.*, 2005). More specifically, before the manipulation task, with the robot set in the initial configuration, the five dots on the object are manually selected in an acquired image. Afterward, tracking of each feature is achieved by processing a region of interest with a size of 80 pixels, centered at the centroid of the dot from the previous image. The feature extraction involved binarization, thresholding, and centroids calculation of the largest blobs with connected pixels in the windows. The extracted trajectories are initially lightly smoothed with a moving average window of three points, and linearly scaled to the length of the longest demonstration (which was equal to 511 frames for the considered experiment). Consequently, the demonstrated input data in the image plane coordinates

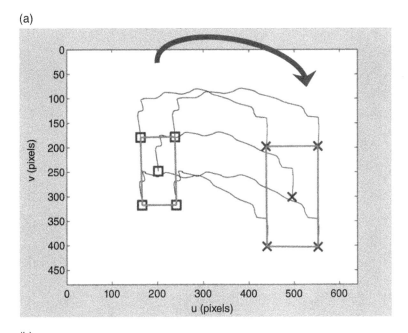

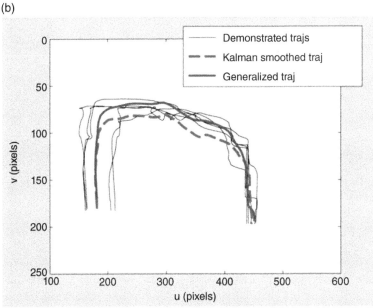

Figure 6.9 (a) Feature trajectories in the image space for one-sample demonstration; (b) demonstrated trajectories, Kalman-smoothed (reference) trajectory, and corresponding planned trajectory for one feature point (for the feature no. 2); (c) demonstrated linear and angular velocities of the object and the reference velocities obtained by Kalman smoothing; (d) Kalman-smoothed (reference) image feature trajectories and the generalized trajectories obtained from the optimization procedure; and (e) demonstrated and the generated Cartesian trajectories of the object in the robot base frame.

(c)

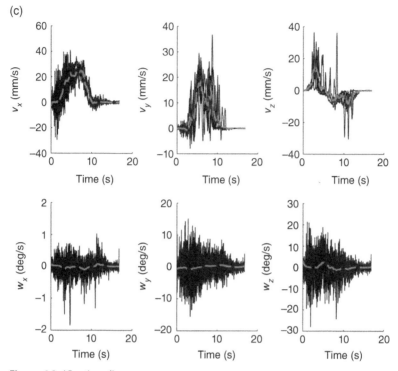

Figure 6.9 (*Continued*)

form the set $\mathbf{r}_m^{(n)}(t_k)$ for $n = 1, \ldots, 5$, $m = 1, \ldots 5$, $k = 1, \ldots, T$. Recall that, in the figures, the image features are shown in pixel coordinates $\mathbf{u}_m^{(n)}(t_k)$, for more intuitive presentation.

For the Kalman smoothing, the measurement and process noise covariance matrices corresponding to the image feature parameters are $\boldsymbol{\Sigma}_{\mathbf{r}}^n = 100\mathbf{I}_{(4 \times 4)}$ and $\mathbf{Y}_{\mathbf{r}}^n = 100\mathbf{I}_{(4M \times 4M)}$, respectively. For the object's pose, the corresponding matrices are $\boldsymbol{\Sigma}_{\mathbf{P}_o} = 0.1\mathbf{I}_{(6 \times 6)}$, $\mathbf{Y}_{\mathbf{P}_o} = 10\mathbf{I}_{(6M \times 6M)}$, and for the object's velocities are $\boldsymbol{\Sigma}_{\mathbf{v}_o} = 0.1\mathbf{I}_{(6 \times 6)}$, $\mathbf{Y}_{\mathbf{P}_o} = 10\mathbf{I}_{(6M \times 6M)}$. The initialization based on the given parameters imposes greater smoothing on the Cartesian position and velocity data, since these parameters are calculated indirectly from the noisy image measurements (by using planar homography transformation). The demonstrated and the reference velocities are shown in Figure 6.9c. On the other hand, the selected initialization for the image feature trajectories results in tight following of the demonstrated trajectories in the image space. The smooth average trajectory

(d)

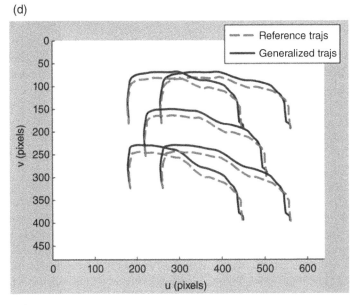

(e)

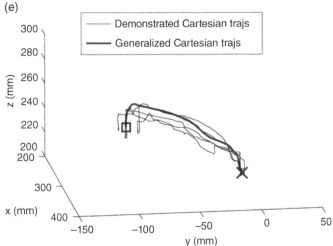

Figure 6.9 (*Continued*)

from the Kalman smoothing for the feature point 2 (i.e., $\mathbf{u}^{(2),\text{ref}}$) is depicted by the thick dashed line in Figure 6.9b.

Afterward, the optimization procedure described in Section 6.4 is performed. The discrete sampling period is set to $\Delta t_k = 0.033$ seconds (which corresponds to the camera frame rate), for $k = 1, 2, ..., 511$. The adopted values for the weighting coefficients, $\{\alpha_n\}_{n=1}^5 = 0.1$ and $\alpha_\upsilon = \alpha_\omega = 0.5$, assign higher weights on the importance for emulating the reference velocities. Less emphasis is placed on following the reference image trajectories. The overall computational time for solving the optimization procedures for this experiment was approximately 30 seconds. The output trajectory from the optimization step is shown in pixel coordinates in Figure 6.9b for the feature point 2 with a thick solid line, superimposed with the demonstrated trajectories and the reference trajectory. Additionally, the output image trajectories for all five feature points are shown in Figure 6.9d, and the corresponding Cartesian trajectory of the object expressed in the robot base frame is shown in Figure 6.9e.

In Figure 6.9b, it is noticeable that the reference trajectory obtained by the Kalman smoother (the dashed line) does not correspond to the geometric mean of the demonstrations, and at some points, it does not belong to the envelope of the demonstrated trajectories. The reason behind this is the temporal variations across the trajectories due to the random distributions of velocities in the demonstrations, that is, the trajectories are linearly scaled to an equal length, but they are not temporally scaled. Therefore, the observed features locations at particular time instants t_k correspond to different states (i.e., spatial positions) across the demonstrated set of M trajectories. On the other hand, it can be noticed that the generated trajectory from the second-order conic optimization is constrained to lie within the envelope of the demonstrated trajectories. One could note here that if the demonstrated trajectories were temporally scaled (for instance, by using the DTW algorithm described in Section 3.3.2), the resulting reference trajectories of the features points would be bounded within the demonstrated envelope, thus they would be better representatives of the task. However, the disadvantage of this approach is the distortion of the velocity profile of the demonstrations caused by the temporal warping, and therefore, it is not considered for the problem at hand.

To move the object by the robot along the generalized trajectory from the optimization model, the IBVS tracking control from (6.58) is employed. The control is implemented through the QuaRC open-architecture system with a rate of 33 milliseconds, while the internal loop for low-level control of the joint motors operates at 1 millisecond rate. The resulting trajectories of the object features during the task execution by the robot are shown in Figure 6.10a (the line with circle markers). For comparison, the desired trajectories from the optimization process are also shown in the figure, depicted with thick solid lines. Regarding the selection of the control gain λ, recall that higher gains dictate fast adaptation to the desired trajectory accompanied with reduced accuracy due to

(a)

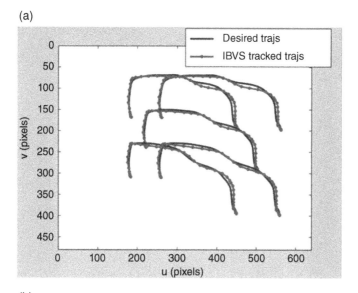

(b)

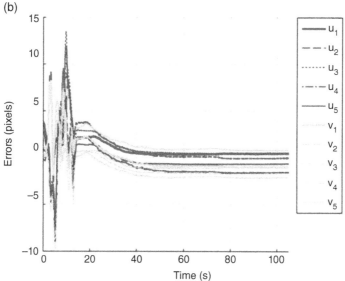

Figure 6.10 (a) Desired and robot-executed feature trajectories in the image space; (b) tracking errors for the pixel coordinates (u, v) of the five image features in the image space; (c) tracking errors for x-, y-, and z-coordinates of the object in the Cartesian space; (d) translational velocities of the object from the IBVS tracker.

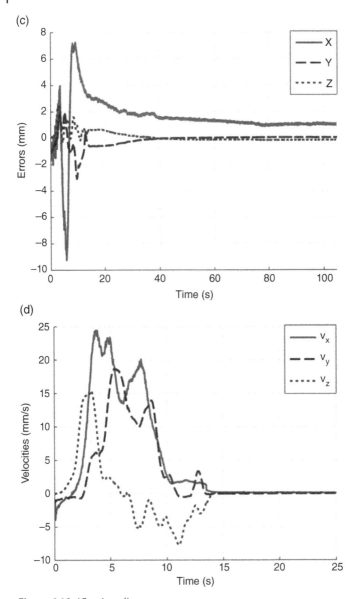

Figure 6.10 (*Continued*)

overshoots, and vice versa for the lower control gains. Also, setting the control gain too high can destabilize the system. Thus, initially the performance of the system was evaluated for different values of the control gain. Subsequently, for the parameter λ, the value of 0.75 was adopted for the first 4/5th of the trajectory length—in order to ensure accurate tracking, and the value of 0.1 was adopted for the last 1/5th of the trajectory—to provide accurate positioning at the end of the task. The tracking errors in the image plane are shown in Figure 6.10b, for the horizontal (u) and vertical (v) pixel coordinates of the five features. The errors during the trajectory following are moderate, with values in the range of several pixels, and with the errors at the final position settling within 2 pixels. The tracking errors in the Cartesian space are shown in Figure 6.10c. The steady-state errors at the final object position expressed in the robot base frame are (1.05, 0.04, −0.15) millimeters. The x-coordinate corresponds to the depth measurements in the camera frame. As expected, feature depths are more sensitive to image noise, due to the camera projection model. The command values for the linear velocities of the object are shown in Figure 6.10d.

The presented approach is important because it establishes a framework for exploiting the robustness of vision-based control in PbD learning. For instance, consider the trajectory tracking results shown in Figure 6.11, which correspond to the same object manipulation task, but this time the reference trajectories from the Kalman smoother are used as desired trajectory for the IBVS tracking. In other words, the optimization procedure was not performed. The results indicate that the tracking is less successful (Figure 6.11a), yielding steady-state image-space errors of approximately 5 pixels (Figure 6.11b). Moreover, the object position at the end of the task (Figure 6.11c) implies divergence from the desired position, as well as high command velocities at the beginning of the tracking (Figure 6.11d). These results reveal the difficulties associated with the trajectory planning in the image space due to nonlinearity in the mapping between the image space and the Cartesian space. More specifically, the positions of the five feature points in the image plane may not necessarily correspond to a realizable Cartesian pose of the object, and/or they may impose large velocities of the object in the Cartesian space. In the considered case, the final positions of the image feature parameters are not achievable, that is, there exist steady-state errors in the image space that the control scheme attempts to minimize, which result in the task failure with regards to the final Cartesian position of the object (Figure 6.11c).

6.5.2.2 Experiment 2

The second experiment is designed to evaluate the performance of the presented approach for trajectories with sharp turns, with the results presented in Figure 6.12. The task is repeated four times by a human teacher via kinesthetic demonstrations. The trajectory learning is performed using the same learning parameters as in Experiment 1. The robot-executed trajectory shown in Figure 6.12a indicates overshoots of the

(a)

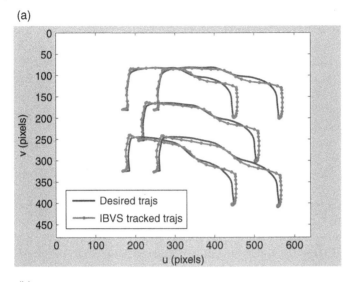

(b)

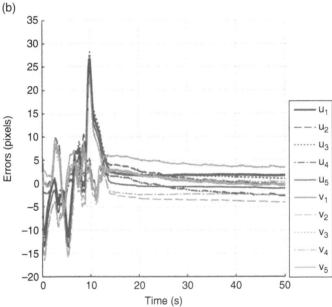

Figure 6.11 Task execution without optimization of the trajectories: (a) Desired and robot-executed feature trajectories in the image space; (b) tracking errors for the pixel coordinates (u, v) of the five image features in the image space; (c) tracking errors for x-, y-, and z-coordinates of the object in the Cartesian space; (d) translational velocities of the object from the IBVS tracker.

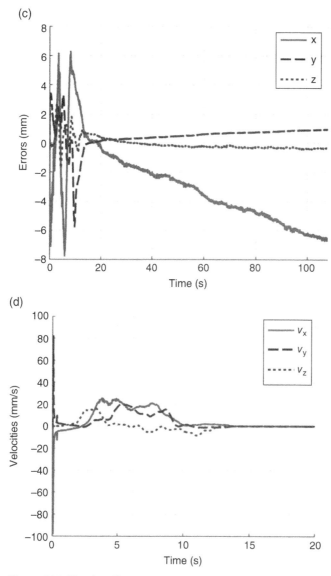

Figure 6.11 (*Continued*)

desired feature trajectories at the two corners where the sharp turns occur. If the control gain λ was set to lower values, the overshoots could have been avoided but on account of cutting the corners. Figure 6.12b shows the task reproduction by the robot, in the case when the generated desired trajectory is interpolated to a trajectory three times longer than

(a)

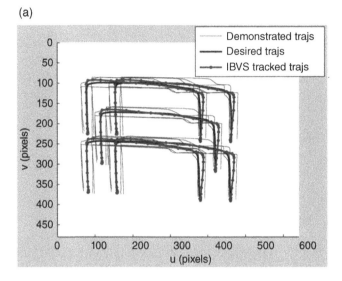

(b)

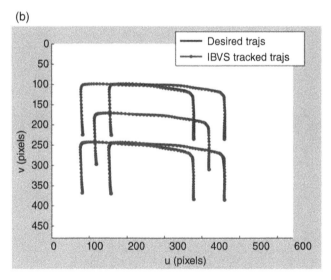

Figure 6.12 (a) Demonstrated trajectories of the image feature points, superimposed with the desired and robot-executed feature trajectories; (b) desired and executed trajectories for slowed down trajectories.

the initial trajectory, that is, this corresponds to slowing down the trajectory three times. In this case, the trajectory following is performed without deviations from the desired trajectory. It can be concluded that the visual tracking performs better for slower trajectories, while fast changing trajectories produce higher tracking errors.

6.5.2.3 Experiment 3

The third experiment is designed to evaluate the tasks involving trajectories with intersecting parts, which is often a challenging case for learning (Khansari-Zadeh and Billard, 2011). A loop in the demonstrated task
is shown in the image plane trajectories in Figure 6.13a. The results from
the trajectory planning step with four demonstrations are displayed in
Figure 6.13b. Similar to the case with sharp turns in Experiment 2, the
presence of the loop caused tracking errors for that part of the trajectory,
and therefore the desired trajectory was interpolated to a three times
longer trajectory. The result of the tracking in this case is given in
Figure 6.13c. The image plane steady-state errors of the IBVS tracking
are within 1.5 pixels (Figure 6.13e), while the translational steady-state
errors of the object in the Cartesian space are (2.4, –0.3, –0.1) millimeters
(Figure 6.13f). The results from the task planning and execution steps are
satisfactory with small trajectory following errors and small positioning
errors of the object at the end of the task.

6.5.3 Robustness Analysis and Comparisons with Other Methods

The advantages of image-based control stemming from the robustness to
modeling errors are next evaluated for several types of uncertainties. First,
the outcome of the learning from demonstration process is examined
under the assumption that the camera is not properly calibrated. The
errors in the intrinsic camera parameters ε ranging from 5 to 80%
are introduced for all considered intrinsic parameters as follows:
$u_0 = (1 + \varepsilon)n_0$, $v_0 = (1 + \varepsilon)v_0$, $f_C k_u = (1 + \varepsilon)f_C k_u$, and $f_C k_v = (1 + \varepsilon)f_C k_v$.
Obviously, these calibration errors affect the transformation in (2.1), that
is, the transformation of the measurements from pixel coordinates into
the spatial image plane coordinates. In particular, the estimation of the
object pose from image coordinates obtained under camera model errors
produces erroneous results. In image-based tracking, the control law
operates directly over parameters in the image space and compensates
for these errors.

The robot performance is evaluated using the optical motion capture
system Optotrak (Section 2.1). Optical markers are attached onto the
object (see Figures 6.4 and 6.7), enabling high accuracy in the measurements of object's poses over time, in comparison to the estimated poses
from camera-acquired images.

The following comparison parameters are used for quantifying the
deviation between the demonstrated trajectories and the robot-executed
trajectories:

1) Steady-state errors of the object's position: $\left|\bar{\xi}_{ss} - \xi_{ss}\right|$, where for each
   coordinate $\xi \in \{x, y, z\}$, ξ_{ss} denotes the steady-state position of the

(a)

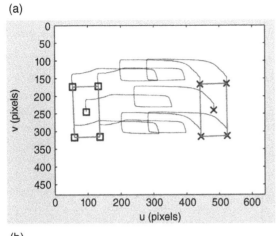

(b)

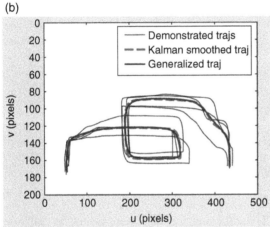

(c)

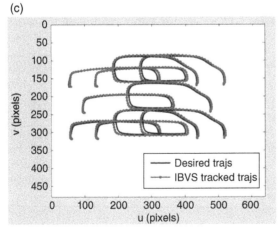

Figure 6.13 (a) Image feature trajectories for one of the demonstrations; (b) demonstrated trajectories, reference trajectory from the Kalman smoothing, and the corresponding generalized trajectory for one of the feature points; (c) desired and robot-executed image feature trajectories; (d) translational velocities of the object from the IBVS tracker; (e) tracking errors for pixel coordinates (u, v) of the five image features in the image space; and (f) tracking errors for x-, y-, and z-coordinates of the object in the Cartesian space.

(d)

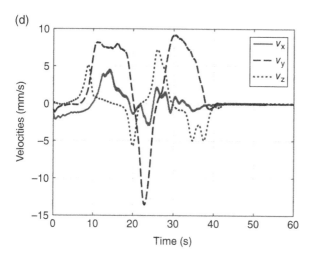

(e)

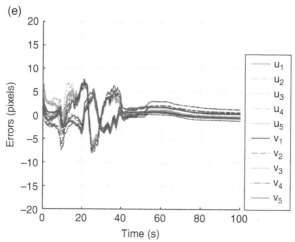

(f)

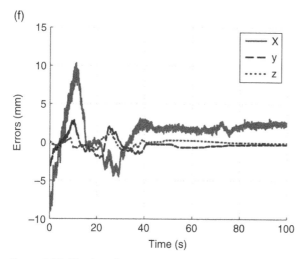

Figure 6.13 (*Continued*)

robot-executed trajectory, and $\bar{\xi}_{ss} = \left(\sum_{m=1}^{M} \xi_{ss}^{(m)}\right)/M$ denotes the mean steady-state position from the set of M demonstrated trajectories

2) RMS deviations for each coordinate $\xi \in \{x, y, z\}$

$$
\vartheta_{\xi} = \left[\frac{1}{M}\sum_{m=1}^{M}\frac{1}{T_{\text{gen}}}\sum_{k=1}^{T_{\text{gen}}}\left(\xi(t_k) - \xi^{(m)}(t_k)\right)^2\right]^{1/2}
\tag{6.61}
$$

3) Overall RMS deviations, obtained by summing the deviations for all three coordinates

$$
\vartheta_{\text{total}} = \left[\sum_{\xi=x,y,z}\frac{1}{M}\sum_{m=1}^{M}\frac{1}{T_{\text{gen}}}\sum_{k=1}^{T_{\text{gen}}}\left(\xi(t_k) - \xi^{(m)}(t_k)\right)^2\right]^{1/2}
\tag{6.62}
$$

For the IBVS tracking, the values of the parameters are shown in Table 6.1. The results demonstrate the robustness of the IBVS scheme, since even for very large intrinsic errors, for example, 80%, the performance of the tracking is almost unchanged. The errors in the object's position at the end of the task are in submillimeter amounts for the y- and z-coordinates. For the x-coordinate of object's position, the errors are larger, since the x-axis corresponds to the depth coordinate of the camera. Even with a well-calibrated camera (i.e., 0% intrinsic errors in the table), the reference values of the steady-state errors are approximately 8 millimeters. When errors in the intrinsic camera parameters

Table 6.1 Evaluation of the trajectory tracking with IBVS under intrinsic camera parameters errors ranging from 0 to 80%.

Intrinsic errors (%)	Steady-state errors			RMS errors			Total RMS errors
	x	y	z	x	y	z	
0	8.2	0.2	0.4	11.1	9.3	5.3	15.4
5	8.3	0.1	0.5	10.8	8.8	4.8	14.7
10	8.3	0.04	0.5	9.4	7.8	3.8	12.8
20	8.6	0.02	0.7	10.5	8.3	4.1	13.9
40	8.4	0.1	0.6	10.3	7.6	3.5	13.3
80	10.6	0.6	0.7	18.7	12.1	6.4	23.2

Tracking errors are expressed in millimeters.

are introduced, the positioning of the object at the end of the task is almost unchanged. Similarly, the tracking errors, quantified with the RMS metrics in Table 6.1, indicate that the trajectory following under intrinsic camera errors is comparable to the performance without introducing intrinsic camera errors. Moreover, the total RMS deviations reveal slightly lower deviations for the case when 10% intrinsic errors are introduced, which may be attributed to the random noise in the image measurements.

For comparison, the task execution is repeated without vision-based control, that is, the robot uses its motor encoder readings for following a desired trajectory in order to relocate the object. The demonstrated object trajectories in the Cartesian space are extracted from the camera-sourced images. Two PbD methods are employed for generating a generalized trajectory: (i) Gaussian mixture model/Gaussian mixture regression (GMM/GMR) approach (Calinon, 2009) with eight mixtures of Gaussians used for modeling the demonstrated object positions, and (ii) dynamic motion primitives (DMPs) approach (Ijspeert *et al.*, 2003) with 20 Gaussian kernels used. Velocities of the object in the Cartesian space are used as command inputs for the robot in order to move the object along the generalized trajectories. Subsequently, an operational space velocity controller for the robot's motion is employed in both cases, based on an inverse differential kinematics scheme (Sciavicco and Siciliano, 2005). The deviations between the demonstrated trajectories and the robot-executed trajectories measured by the optical tracker are given in Table 6.2. The results indicate deteriorated performance, with high tracking errors, as well as errors in the object positioning at the end of the task.

Table 6.2 Evaluation of the trajectory tracking without vision-based control under intrinsic camera parameters errors ranging from 0 to 80%.

Intrinsic errors (%)	GMM/GMR approach				DMPs approach			
	Steady-state errors			Total RMS errors	Steady-state errors			Total RMS errors
	X	y	Z		x	Y	z	
0	6.5	3.4	17.0	17.7	4.2	1.5	13.3	18.6
5	11.7	5.0	18.2	19.1	10.2	3.0	14.8	20.3
10	15.9	6.8	20.2	26.5	15.7	5.2	16.4	23.4
20	22.3	9.62	22.5	29.9	28.5	9.6	19.6	32.4
40	33.5	14.3	25.3	36.8	47.1	16.3	25.8	39.8
80	46.2	30.4	35.4	43.9	68.6	29.4	36.7	64.9

Tracking errors are expressed in millimeters.

The deviations of the robot-executed trajectories from the demonstrated trajectories originate from two sources: the object pose estimation using erroneous camera parameters and errors due to robot kinematics. Namely, even for the case without errors in the intrinsic camera parameters (first row in Table 6.2 with 0% errors), there exist tracking and steady-state position errors due to the robot kinematics (such as incorrect estimate of the robot's end-point position after the homing, coupling of the robot's joints, backlashes, and frictions in the actuators). On the other hand, by using visual feedback to control the robot directly in the image space of the camera, the errors associated with the inaccuracies in the robot kinematic model are avoided.

Robustness to extrinsic camera parameters is also evaluated in a similar fashion. Recall that the extrinsic parameters refer to the pose of the camera with respect to a given frame. The presented vision-based tracker is very robust to the extrinsic camera errors, since these parameters are hardly used in the control law (6.58), with exception of the rotation matrix R_c^b. The obtained results for the robot performance are almost identical for errors ranging from 5 to 80%, and, therefore, are not reported here. The GMM/GMR and DMPs approaches were also examined in the case of adding extrinsic camera errors. As expected, the camera pose errors induced erroneous estimation of the object pose from grabbed images, that is, the object trajectories with respect to the robot base are translated into the 3D space. However, the performance of both methods was not affected in this case, due to the type of employed robot control scheme. More precisely, the used velocity controller is independent of the initial position of the robot's end point. Therefore, the object was moved along the desired trajectories without any undesired effects caused by the errors in the extrinsic camera parameters. On the contrary, if a position controller was used for moving the robot's gripper (and the object) along the planned trajectories, the executed robot trajectory would have differed significantly from the planned trajectory. For comparison, the mean values of the final object pose from the demonstrated trajectories, extracted under assumption of the presence of camera extrinsic errors, are shown in Table 6.3.

Reproduction of learned tasks by using visual servoing can also be implemented by first performing the task planning in the Cartesian space, then projecting several salient features of the target object onto the image space, and lastly employing image-based tracker for following the image feature trajectories (Mezouar and Chaumette, 2002). This scenario assumes independent planning and execution steps, and ensures robust execution under uncertainties, due to the implemented vision-based tracking. However, in the case of modeling errors,

Table 6.3 Coordinates of the object position at the end of the task expressed in the robot base frame (in millimeters) under extrinsic camera parameters errors ranging from 0 to 80%.

Extrinsic errors (%)	X	y	z
0	351.5	−9.1	214.9
10	416.6	−12.8	238.5
40	611.7	−23.9	309.1
80	871.9	−38.9	403.23

projections of the features from the Cartesian space onto the image plane can result in robot actions that differ from the planned ones. Figure 6.14a shows the projected image plane trajectories from the planning step with a thick continuous line. The projected trajectories in the cases of intrinsic camera errors ε of 5, 10, and 20% applied to the camera intrinsic parameters $(u_0 = (1 + \varepsilon)u_0, \ v_0 = (1 + \varepsilon)v_0,$ $f_C k_u = (1 + \varepsilon)f_C k_u, f_C k_v = (1 + \varepsilon)f_C k_v)$ are shown overlaid in Figure 6.14a. Hence, mapping from the Cartesian into the image space under modeling errors causes deviation from the planned trajectories. Due to the displacement of the principal point coordinates for the case of 20% errors, the image feature trajectories are out of the FoV of the camera. Figure 6.14b presents another case when moderate 5% errors have been imposed only on the intrinsic parameters corresponding to the focal length scaling factors $f_C k_u$ and $f_C k_v$. The plot illustrates that the projected trajectories are elongated and, as expected, they differ from the planned ones. These deviations of the projected trajectories will result in robot-executed motions that are different from the planned motions. On the other hand, the approach presented here uses the camera-observed feature trajectories both for task planning and for task reproduction, thus avoiding the problems with the 2D–3D mapping, and providing a learning framework robust to camera modeling errors.

6.6 Discussion

For generating the reference trajectories, instead of Kalman smoothing, other machine learning methods can be employed, for example, HMM (Chapter 4), GMM/GMR (Calinon, 2009), and dynamical systems approach (Ijspeert et al., 2003). However, since the reference trajectories are subjected to correction for visual tracking (through the described

(a)

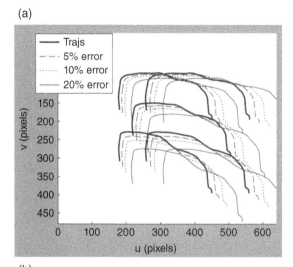

Figure 6.14 Projected trajectories of the feature points in the image space with (a) errors of 5, 10, and 20% introduced for all camera intrinsic parameters; (b) errors of 5% introduced for the focal length scaling factors of the camera.

(b)

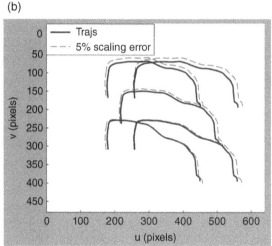

optimization procedure), a simple averaging via Kalman smoothing is adopted. Although the outputs of the Kalman smoothing depend on the initialization, it is a computationally fast algorithm, and can work for trajectories with different shape complexity (Coates *et al.*, 2008).

Among the shortcomings and possible improvements of the presented approach against the existing methods are (i) the controlled system is locally stable; however, it is difficult to theoretically derive the region of stability; (ii) the implemented convex optimization in the planning phase warrants global convergence within the feasible solutions; however, if the constraints are set too strict, the obtained optimal solution might not be satisfactory; (iii) the visual tracking can produce large tracking errors for high-speed tracking tasks; (iv) deriving a generic performance metrics with regards to the optimality of the generated solution for task reproduction is challenging and remains to be done; (v) the demonstrated trajectories are linearly scaled, which can result in poor generalization, especially when the demonstrations vary significantly in length and velocity; and (vi) the object geometry is assumed to be known; however, in a general case, it is preferred to consider unknown object geometry.

6.7 Summary

In this chapter, a method for learning from demonstrations based on integration of a visual controller for robust task execution is presented. The method provides a unified framework for performing the perception, planning, and execution steps in the image space of a vision sensor. It enables to perform the task planning and execution directly over the observed parameters, that is, the projections of several features on the target object onto the vision sensor space. Such direct mapping endows robustness to errors originating from the calibration of the perception sensor, measurement noise, and the robot kinematics.

The task planning is solved in two substeps. First, a machine learning approach is employed to generate a reference trajectory for task reproduction from multiple examples of the task. Afterward, at each sampling time, an optimization procedure is performed, which is initialized by the reference trajectory, and aims to modify the reference trajectory in order to enforce satisfaction of task constraints. The formalization of the optimization problem involves vision sensor constraints (related to the image-plane limits and the bounds of the demonstrated image trajectories), 3D Cartesian constraints (related to the bounds of the demonstrated workspace and the velocity limits), and robot constraints (related to the joint saturations and robot kinematics). The robustness of the method is evaluated through a set of simulations and experiments conducted for learning and reproduction of trajectories with different complexities.

References

Asada, M., Yoshikawa, Y., and Hososda, K., (2000). Learning by observation without three-dimensional reconstruction. *Proceedings of Sixth International Conference on Intelligent Autonomous Systems*, Venice, Italy, pp. 555–560.

Bouguet, J.-I., (2010). *Camera Calibration Toolbox for MATLAB*. Available from: http://www.vision.caltech.edu/bouguetj/calib_doc/index.html (accessed on September 9, 2016).

Calinon, S., (2009). *Robot Programming by Demonstration: A Probabilistic Approach*. Boca Raton, USA: EPFL/CRC Press.

Cervera, E., (2003). *Visual Servoing Toolbox for MATLAB/Simulink*. Available from: http://vstoolbox.sourceforge.net/ (accessed on September 9, 2016).

Chaumette, F., (1998). Potential problems of stability and convergence in image-based and position-based visual servoing. *The Confluence of Vision and Control*. (Eds.) Kriegman, D., Hager, G., and Morse, A.S. Lecture Notes in Control and Information Systems, No. 237. London, UK: Springer-Verlag, pp. 66–78.

Chaumette, F., and Hutchinson, S., (2006). Visual servo control—Part I: Basic approaches. *IEEE Robotics and Automation Magazine*, vol. **13**, no. 4, pp. 82–90.

Chaumette, F., and Hutchinson, S., (2007). Visual servo control—Part II: Advanced approaches. *IEEE Robotics and Automation Magazine*, vol. **14**, no. 1, pp. 109–118.

Chesi, G., and Hung, Y.S., (2007). Global path-planning for constrained and optimal visual servoing. *IEEE Transactions on Robotics*, vol. **23**, no. 5, pp. 1050–1060.

Coates, A., Abbeel, P., and Ng, A.Y., (2008). Learning for control from multiple demonstrations. *Proceedings of International Conference on Machine Learning*, Helsinki, Finland, pp. 144–151.

Corke, P., (2011). Robotics, vision and control: fundamental algorithms in MATLAB. *Springer Tract in Advanced Robotics*, vol. **73**. (Eds.) Siciliano, B., Khatib, O., and Groen, F., Berlin, Germany: Springer-Verlag.

Corke, P., and Hutchinson, S., (2001). A new partitioned approach to image-based visual servo control. *IEEE Transactions on Robotics and Automation*, vol. **17**, no. 4, pp. 507–515.

Deng, L., Janabi-Sharifi, F., and Wilson, W., (2005). Hybrid motion control and planning strategies for visual servoing. *IEEE Transactions on Industrial Electronics*, vol. **52**, no. 4, pp. 1024–1040.

Dillmann, R., (2004). Teaching and learning of robot tasks via observation of human performance. *Robotics and Autonomous Systems*, vol. **47**, no. 2–3, pp. 109–116.

Ficocelli, M., and Janabi-Sharifi, F., (2001). Adaptive filtering for pose estimation in visual servoing. *Proceedings of IEEE/RSJ International Conference on Intelligent Robots and Systems*, Maui, USA, pp. 19–24.

Gans, N.R., and Hutchinson, S.A., (2007). Stable visual servoing through hybrid switched-system control. *IEEE Transactions on Robotics*, vol. **23**, no. 3, pp. 530–540.

Hersch, M., Guenter, F., Calinon, S., and Billard, A., (2008). Dynamical system modulation for robot learning via kinesthetic demonstrations. *IEEE Transactions on Robotics*, vol. **24**, no. 6, pp. 1463–1467.

Horaud, R., Canio, B., Leboullenx, O., and Lacolle, B., (1989). An analytic solution for the perspective 4-point problem. *Computer Vision, Graphics, and Image Processing*, vol. **47**, pp. 33–44.

Hutchinson, S., Hager, G., and Corke, P., (1996). A tutorial on visual servo control. *IEEE Transactions on Robotics and Automation*, vol. **12**, no. 5, pp. 651–670.

Ijspeert, A.J., Nakanishi, J., and Schaal, S., (2003). Learning attractor landscapes for learning motor primitives. *Advances in Neural Information Processing Systems*, vol. **15** (Eds.) Becker, S., Thrun, S., and Obermayer, K., Cambridge, USA: MIT Press, pp. 1547–1554.

Janabi-Sharifi, F., (2002) Visual servoing: theory and applications. *Opto-Mechatronics Systems Handbook*. (Ed.) Cho, H., Boca Raton, USA: CRC Press, ch. 15.

Jiang, P., and Unbehauen, R., (2002). Robot visual servoing with iterative learning control. *IEEE Transactions on Systems, Man, and Cybernetics—Part A*, vol. **32**, no. 2, pp. 281–287.

Jiang, P., Bamforth, L.C.A., Feng, Z., Baruch, J.E.F., and Chen, Y.Q., (2007). Indirect iterative learning control for a discrete visual servo without a camera-robot model. *IEEE Transactions on Systems, Man, and Cybernetics—Part B*, vol. **37**, no. 4, pp. 863–876.

Khansari-Zadeh, S.M., and Billard, A., (2011). Learning stable non-linear dynamical systems with Gaussian mixture models. *IEEE Transactions on Robotics*, vol. **27**, no. 5, pp. 943–957.

Kormushev, P., Nenchev, D.N., Calinon, S., and Caldwell, D.G., (2011). Upper-body kinesthetic teaching of a free standing humanoid robot. *Proceedings of International Conference on Robotics and Automation*, Shangai, China, pp. 3970–3975.

Liu, Y., Huang, T.S., and Faugeras, O.D., (1990). Determination of camera location from 2-D to 3-D line and point correspondences. *IEEE Transactions on Pattern Analysis and Machine Intelligence*, vol. **12**, no. 1, pp. 28–37.

Malis, E., Chaumette, F., and Boudet, S., (1999). 2-1/2-D visual servoing. *IEEE Transactions on Robotics and Automation*, vol. **15**, no. 2, pp. 238–250.

Marchand, E., Spindler, F., and Chaumette, F., (2005). VISP for Visual Servoing: a generic software platform with a wide class of robot control skills. *IEEE Robotics and Automation Magazine*, vol. **12**, no. 4, pp. 40–52.

Mezouar, Y., and Chaumette, F., (2002). Path planning for robust image-based control. *IEEE Transactions on Robotics and Automation*, vol. **18**, no. 4, pp. 534–549.

Nematollahi, E., Vakanski, A., and Janabi-Sharifi, F., (2012). A second-order conic optimization-based method for visual servoing. *Journal of Mechatronics*, vol. **22**, no. 4, pp. 444–467.

QuaRC. *Accelerate Design*, Quanser, (2012). Available from: http://www. quanser.com (accessed on September 9, 2016).

Rauch, H.E., Tung, F., and Striebel, C.T., (1965). Maximum likelihood estimates of linear dynamic systems. *Journal of the American Institute of Aeronautics and Astronautics*, vol. **3**, no. 8, pp. 1445–1450.

Sciavicco, L., and Siciliano, B., (2005). *Modeling and Control of Robot Manipulators* (second edition). London, UK: Springer-Verlag.

Shademan, A., Farahmand, A.-M., and Jagersand, M., (2009). Towards learning robotics reaching and pointing: an uncalibrated visual servoing approach. *Proceedings of Canadian Conference on Computer and Robot Vision*, Kelowna, Canada, pp. 229–236.

Sturm, J.F., (1997). *SeDuMi 1.21, a MATLAB Toolbox for Optimization over Symmetric Cones*. Available from: http://sedumi.ie.lehigh.edu/ (accessed on September 9, 2016).

Vakanski, A., Janabi-Sharifi, F., Mantegh, I., and Irish, A., (2010). Trajectory learning based on conditional random fields for robot programming by demonstration. *Proceedings of IASTED International Conference on Robotics and Applications*, Cambridge, USA, pp. 401–408.

Wilson, W.J., Hulls, C.C., and Janabi-Sharifi, F., (2000). Robust image processing and position-based visual servoing. *Robust Vision for Vision-Based Control of Motion*. (Eds.) Vincze, M., and Hager, G.D., New Jersey: IEEE Press, pp. 163–201.

Yuan, J.S.-C., (1989). A general photogrammetric method for determining object position and orientation. *IEEE Transactions on Robotics and Automation*, vol. **5**, no. 2, pp. 129–142.

Index

Robot Learning by Visual Observation, First Edition. Aleksandar Vakanski and
Farrokh Janabi-Sharifi.
© 2017 John Wiley & Sons, Inc. Published 2017 by John Wiley & Sons, Inc.